BASIC NUCLEAR MEDICINE

APPLETON-CENTURY-CROFTS/New York
A Publishing Division of Prentice-Hall, Inc.

BASIC NUCLEAR MEDICINE

Sheldon Baum, M.D.

Associate Professor of Radiology (Nuclear Medicine) and Assistant Professor of Medicine, New York Medical College, New York, N.Y.

Roland Bramlet, Ph.D.

Chief Physicist, Highland Hospital, and Consulting Physicist, Rochester General Hospital, Rochester, N.Y.

Library of Congress Cataloging in Publication Data

Baum, Sheldon.
Basic nuclear medicine.

Bibliography: p.
1. Radiology, Medical. I. Bramlet, Roland, joint author. II. Title. [DNLM: 1. Nuclear medicine. WN440 B347b]
RM847.B37 616.07'575 74-23307
ISBN 0-8383-0613-5

75 76 77 78 79 / 10 9 8 7 6 5 4 3 2 1

PRINTED IN THE UNITED STATES OF AMERICA

cover design by Karen Robbins

to my wife
MAUREEN
and
to the memory of my father
ABRAHAM BAUM
1885-1936

S.B.

---

and

to RALPH BADGLEY
teacher and friend

R.B.

PREFACE

Nuclear medicine is now a well-established specialty and is included in the teaching programs of most medical schools and hospitals. For the specialist and others with a particular interest in the field, there are many excellent, detailed monographs and several textbooks that treat the subject exhaustively.

However, for such physicians as residents in radiology and other related specialties, as well as for those physicians already in practice, with little or no prior exposure to nuclear medicine, there is no readily available source of general information in this field to which they may turn without becoming overwhelmed. It is for this large group of physicians that this book is primarily intended. It may also be used by medical students who are engaged in assigned studies or electives in nuclear medicine.

Nuclear medicine becomes interesting for the physician if he is taught, for example, the distribution characteristics of a particular radioactive material when it enters the body and how specific pathologic entities may alter this distribution or uptake. This approach to the clinical aspects of nuclear medicine evolved from the authors' experience in teaching this subject over the past decade and is the method followed in this book. Wherever it seemed useful, the explanation of a concept is amplified with the generous use of diagrams.

There is some departure from the usual treatment of the physical aspects of nuclear medicine, such as the characteristics of radioactive materials and the instruments used in their detection and imaging. The understanding of these subjects becomes more interesting and meaningful when they are presented with special reference to practical analogies and clinical applications. To achieve these objectives the sections on physics and instrumentation related to nuclear medicine are found in the final four chapters of the book, so that the reader might have the benefit of being introduced to specific clinical applications of nuclear medicine procedures before attempting to understand the physical concepts involved.

The aim in writing this book was not to produce a complete text on nuclear medicine. Because this is, in a sense, a primer on nuclear medicine, many aspects and procedures are not covered. For the interested reader who seeks more information in a specific area and for those who may wish to specialize in nuclear medicine, detailed references are suggested at the end of each chapter. It is hoped that the reader will profit from the manner in which the material is presented in this book so that he may develop an understanding of nuclear medicine procedures that is both enlightened and clinically useful.

As every nuclear medicine physician knows, the nuclear medicine technologist occupies a role of unusual responsibility, especially in regard to imaging procedures. Most of the imaging presented in this book was performed by Elaine Harwitz, the supervising nuclear medicine technologist, and Mary Dittman, senior nuclear medicine technologist, at the New York Medical College, Flower and Fifth Avenue Hospital. The contributions of three other nuclear medicine technologists, Mary Louise Joslyn, Pierrette Wise, and Rebecca Gonzalez, are also gratefully acknowledged.

My wife Maureen was of inestimable support in the preparation of the manuscript. Herself an accomplished nuclear medicine technologist, she assisted me in many aspects of sophisticating, simplifying, and detailing radionuclide procedures.

Special thanks are due Dr. Chester W. Fairlie, who introduced me to nuclear medicine and stimulated my interest in the field. Others who made substantial scientific contributions to the efforts involved in writing this book, in some cases many years ago, include Drs. Kevin L. Macken, William T. Taylor, and Richard J. Weber, and Mr. William H. Yates.

It has always been gratifying to receive the assistance, advice, and suggestions of a number of my radiologist colleagues, such as Drs. Farooq P. Agha, Emil J. Balthazar, Steven C. Gurkin, Robert M. Klein, Maria Angelica Mieza, and Natalie S. Strutynsky. Other faculty members who have been helpful in a similar manner include Drs. Jacob L. Brener, Alan B. Rothballer, Sheldon P. Rothenberg, and Robert J. Strobos. Especially appreciated has been the encouragement of Dr. Richard M. Friedenberg, Professor and Chairman, Department of Radiology, New York Medical College.

Thanks are extended to Jo-Ann Terry, who performed most of the stenographic duties associated with the manuscript, and to Diane Alamo. And, of course, David, Robert, and Stephanie.

S.B.
R.B.

CONTENTS

FOREWORD

The availability of a large variety of radioisotopes applicable to the needs of the physician has contributed significantly to the spectacular growth of a modern, exciting, important medical discipline termed nuclear medicine. Radioactive tracer methods have led to a revision of the pathogenesis of certain human diseases, opened new avenues of research, and resulted in new methods of diagnosis and therapy. It is therefore essential that the physician be informed concerning the nature of nuclear energy, its potentialities in the practice of medicine, and its perils when used improperly.

A debt of gratitude is owed to the host of scientists whose talent and creativity gave us the radioisotopes and the instrumentation to use them. Their contributions, in part, underlie the increasingly sophisticated practice of modern medicine. Human society is served by the synergism of all pertinent knowledge and skills. Knowledge cannot be applied to the prevention, diagnosis, and treatment of human ailments until the knowledge exists. The horizons of science continue to expand, and newer and better solutions to the problems that plague all are sought. Complex problems that man could hardly have imagined a generation ago are discerned and solved today. A new biology and a new medicine have burst forth, and the consequences of this intellectual flowering are still to be reckoned. Yet important questions are unanswered—perhaps even unasked.

Seldom in the history of medicine has a medical science made such rapid growth as nuclear medicine. In this fascinating branch of modern medicine, radioactive materials are used in research and in diagnosis and therapy. Nuclear medicine provides the modern physician with additional previously unavailable objective data concerning the architecture and function of cells and organs of the human body. With the proper radioisotope techniques we can determine not only where a radioactive material is in the body but also how much of what kind is there at a given time. Nuclear methods provide us with the methodology to detect clinical changes in form, function, and the biochemistry of disease, as well as its evaluation. Information is added to the data base to make it more comprehensive and thus permit the physician to make a more specific diagnosis. Incorrect data lead to the construction of an erroneous syndrome.

The authors of this new textbook have continually sought to provide a readable, basic, and balanced approach to the subject of nuclear medicine. As indicated in the preface, the text is primarily intended for physicians with little or no previous exposure to the subject matter of nuclear medicine. The book is written in such a fashion as to provide a working knowledge that does not overwhelm the reader. The contents provide basic information concerning the application of tracer techniques in patient management and medical diagnosis.

For more than four decades I have been an avid student and a teacher of physiology in health and disease. Regardless of the blunders, vicissitudes, and avant-garde methods of education, the physiologist and the physician who desire to appreciate fully the human machinery at work must have a first-hand acquaintance with the working parts, ranging from the subatomic level to the total organism.

Learning and teaching should be a great source of pleasure. One of the greatest services a teacher can render is to share with students his love and enthusiasm and his sense of values for the pursuit and understanding of the wondrous phenomenon called life. One of man's noblest endeavors is to express the marvels of the living organism in the language of science.

Probe deeply into the signal triumphs of modern medicine and communicate what you have learned to others. To be a good teacher in any discipline, and this includes nuclear medicine, one must teach with one's life. A prerequisite of being an effective teacher is being a person. Convey a sense of caring about what you are doing and somehow make that concern include your students.

A good physician is certainly a benevolent scientist. Knowledge of science is the catalyst that changes medical diagnosis and treatment to the thinking level. Pettiness of spirit and shabby standards can choke medical progress.

Effective learning and implementation of the sciences are vital aspects of the physician's experience. Let us not ignore the technologic marvels of which medicine is capable, for exotic as they seem at times they are now saving many lives. The physician who ceases to keep abreast of the ever-expanding literature in medicine is no longer a good physician. The education of the doctor of medicine should never terminate.

RICHARD J. WEBER, PH.D.
Professor of Biology
Georgetown University,
Washington, D.C.

INTRODUCTORY NOTE

During the earlier days of nuclear medicine all radionuclide imaging was performed on rectilinear scanners and the images so produced came to be known as scans. Although the images obtained with scintillation cameras do not involve the use of moving detectors, such images nevertheless are often referred to as scans too, or as camera scans.

When viewing a roentgenogram it is customary for the right side of the x-ray to appear to the observer's left, regardless of the position in which the study was performed. Although there is some variance, nuclear medicine images are generally viewed according to the position in which the imaging was obtained. Thus the right side of an anterior scan is on the observer's left, while the right side of a posterior scan is on the observer's right.

BASIC NUCLEAR MEDICINE

PART ONE

Radionuclide Imaging and External Monitoring

chapter 1

CENTRAL NERVOUS SYSTEM

BRAIN IMAGING

Only rarely does the introduction of a new technique radically alter the diagnostic approach to a specific medical problem. However, with the advent of radionuclide brain imaging, the search for space-occupying lesions in patients suspected of having intracranial pathology has been remarkably simplified. The ability to demonstrate brain tumors on scans is one of the major accomplishments of the nuclear medicine era.

Rationale and Radionuclides

The radionuclides used in brain imaging all show evidence of increased concentration or localization in any area of the brain in which there has been a structural alteration, such as a brain tumor or cerebral infarct. The reasons for this increased concentration of radioactive material are not yet clearly understood but are believed to be associated with an alteration in the blood-brain barrier at the diseased site, allowing the tracer to enter this region in relatively large amounts. Thus intracranial disease entities, both neoplastic and nonneoplastic, are reflected on scans as areas of increased radioactivity.

The agent most commonly used today is technetium 99m in the form of sodium pertechnetate. The usual adult dose is 10 to 15 mCi administered intravenously 1 hour before scanning. To block uptake of the pertechnetate into the choroid plexus it is necessary to give the patient potassium perchlorate prior to injection of the ^{99m}Tc. An oral dose of 200 mg given 1 to 8 hours before tracer injection is sufficient. This additional step is essential because scan visualization of the choroid plexus may resemble a cerebral lesion.

The distribution and behavior of the pertechnetate ion (TcO_4^-) in the body are in most instances similar to that of iodide. Pertechnetate is taken up in many glandular structures such as the salivary glands, choroid plexus, and thyroid. In the latter it competes with iodide in regard to trapping but does not become organified. There is some evidence that TcO_4^- becomes protein-bound on entering the bloodstream. Although a principal avenue of pertechnetate excretion is through the kidneys, a significant amount (approximately 40 percent) is secreted in the gastric mucosa and then excreted via the gastrointestinal tract. Because it remains in the gastrointestinal system a relatively long time prior to its excretion by this route, the critical organ in regard to radiation exposure from ^{99m}Tc-sodium pertechnetate is the large intestine.

In the past, other brain scanning agents such as ^{131}I-human serum albumin and ^{197}Hg- and ^{203}Hg-chlormerodrin were commonly employed. They are rarely used for brain scanning today because of their relatively long half-lives and radiation exposure when compared with ^{99m}Tc-sodium pertechnetate as the brain imaging agent.

Normal Brain Imaging

Imaging is routinely performed in anterior, posterior, and both left and right lateral positions. As in other imaging procedures the position of the organ or area being scanned in relation to the detector determines the position of the scan (Fig. 1).

The largest intracranial blood vessels (the large venous sinuses of the dura mater) are normally seen on a brain scan. A normal anterior scan performed on a rectilinear scanner is shown in Figure 2. On the scan the triangular shaped area of increased radioactivity seen centrally is the superior sagittal sinus, with the draining venous lacunae and emissary veins located peripherally. The large, somewhat ovoid areas of absent radioactivity on either side are the cerebral hemispheres. Since they are relatively avascular compared to the surrounding structures, little or no radioactivity is seen in them. The increased radioactivity forming the inferior portion of the anterior

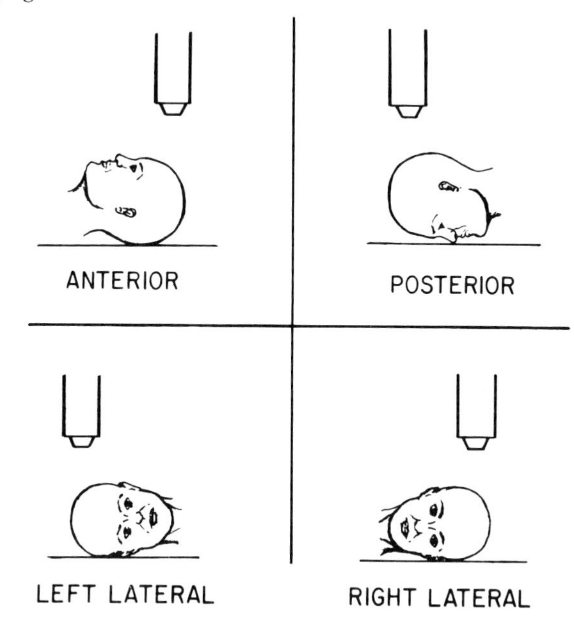

Fig. 1. Four conventional positions for brain imaging. Each position is determined by the relation of the head to the detector.

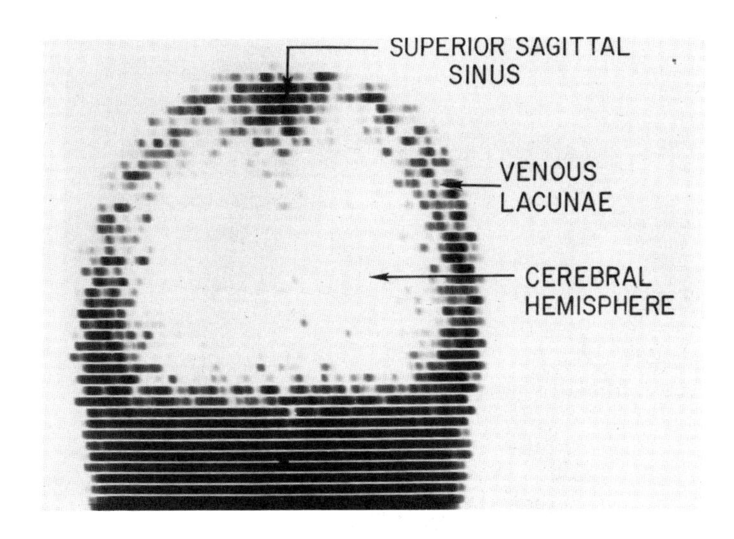

Fig. 2. Normal anterior brain scan.

scan is due to many highly vascular structures including the frontal sinuses, orbits, and nasal mucosa.

The superior sagittal sinus outlines the superior border on the lateral scan (Fig. 3) and ends posteriorly in the bulbous confluence of the sinuses (torcular Herophili). Again, the cerebral hemisphere shows relative absence of radioactivity. The transverse sinus, which runs somewhat anteriorly from the confluence to the start of the sigmoid sinus, serves as a convenient landmark of division between supratentorial and infratentorial (or posterior fossa) structures. For the most part the posterior fossa is the region inferior to the transverse sinus (actually, a tiny portion of the vermis of the cerebellum extends above the level of the transverse sinus). The triangular area of increased radioactivity anteriorly is due to a number of highly vascular structures, including sphenoid, frontal, and maxillary sinuses; cavernous sinus; orbit; and nasal mucosa.

The superior sagittal sinus is seen on the posterior scan (Fig. 4) as it courses downward to enter the confluence of the sinuses. From the confluence, the transverse sinuses extend laterally. As on the lateral scan, the area inferior to the transverse sinuses is the posterior fossa.

It is appropriate at this point to mention briefly the two major methods commonly used for imaging in nuclear medicine. The radionuclide images presented in this book were obtained either on a rectilinear scanner (device with moving detector) or a scintillation camera (device with stationary detector). The scans shown in Figures 2 through 4 were performed on a rectilinear

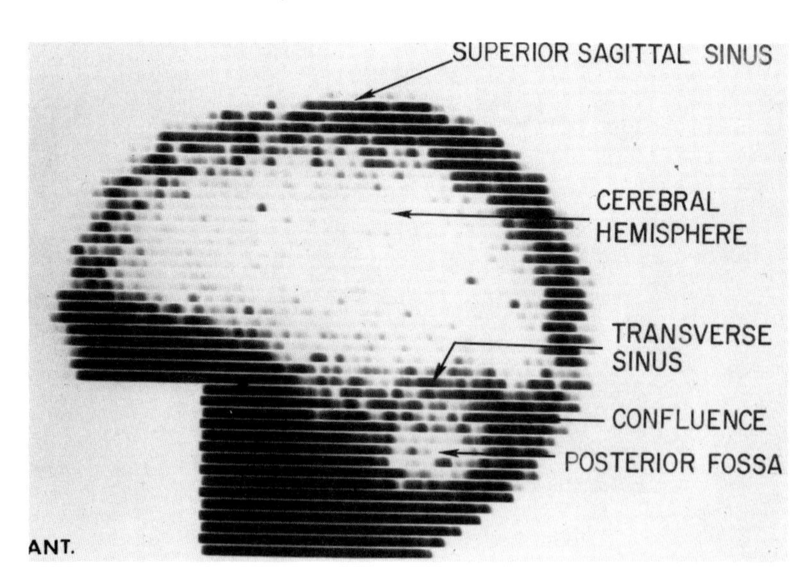

Fig. 3. Normal left lateral brain scan.

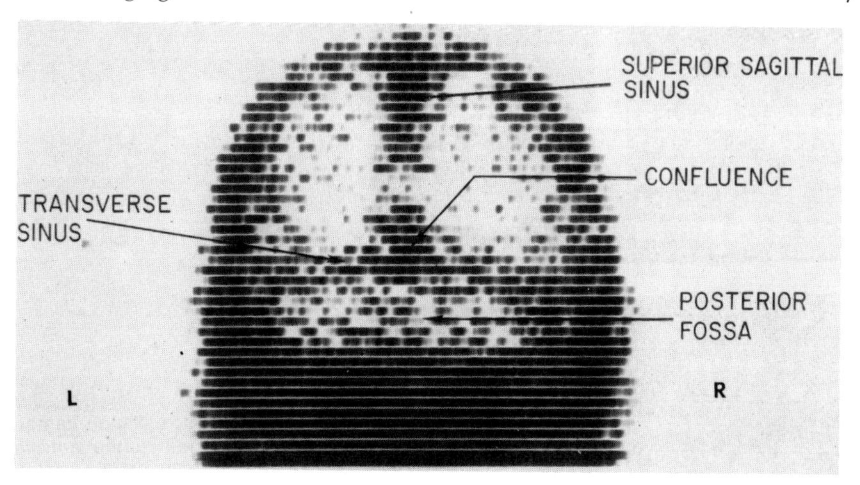

Fig. 4. Normal posterior brain scan.

scanner. Images obtained from a scintillation camera provide similar information and may be presented on Polaroid prints (Fig. 5) or on transparent film (Fig. 6). The characteristics of rectilinear scanners and scintillation cameras are discussed in Chapter 10.

In addition to the four routine scanning positions, a vertex scan is sometimes obtained. The head is hyperextended and the detector placed superiorly (Fig. 7). A normal vertex scan is seen in Figure 8. Imaging in this position may be useful in demonstrating lesions closely related to the midline. To avoid interference from radioactivity in the salivary glands it is helpful to administer atropine prior to giving the radionuclide.

One may readily see that the areas of the brain accessible to radionuclide imaging are for the most part limited to the cerebral hemispheres and posterior cranial fossa. The latter is usually studied for lesions of the cerebellum and cerebellopontine angle.

Brain Tumors

Brain tumors are recognized on scans as localized areas of increased radioactivity. They are usually rounded, discrete, and readily identified.

The reasons a brain tumor may be visualized on a scan are still poorly understood.[1] The presence of the tumor is associated with a breakdown in the blood-brain barrier, this alteration permitting substances from the cerebral circulation to gain access to the neoplasm. The radioactive material enters the region of the tumor and apparently becomes localized or accumulates within the tumor. This localization may occur within the vascular network of

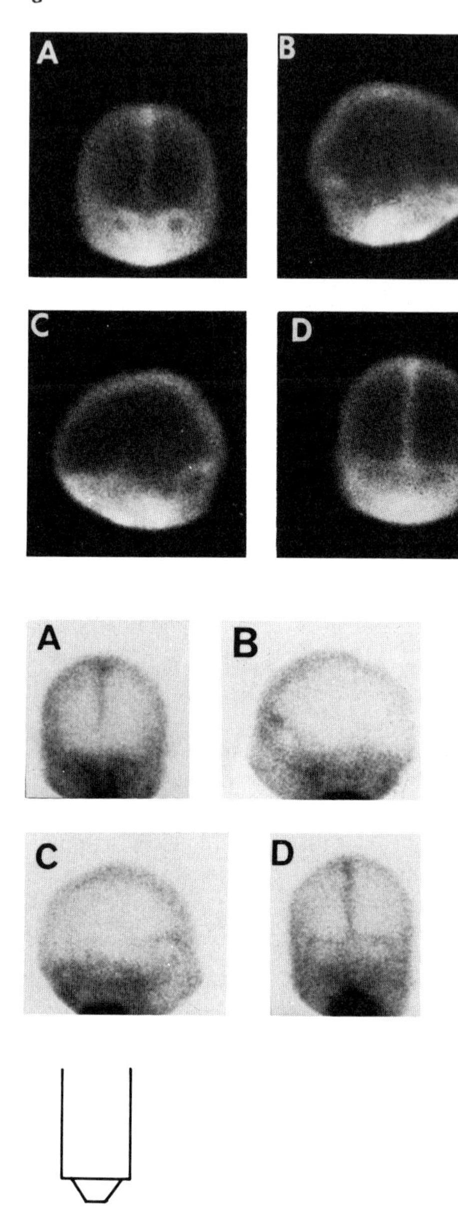

Fig. 5. Normal scintillation camera images performed in the anterior (A), right lateral (B), left lateral (C), and posterior (D) positions. The image is presented on positive photographic paper.

Fig. 6. Normal scintillation camera brain studies (same as in Figure 5), with images presented on transparent film. Positions: A. Anterior. B. Right lateral. C. Left lateral. D. Posterior.

Fig. 7. Vertex position for brain scanning.

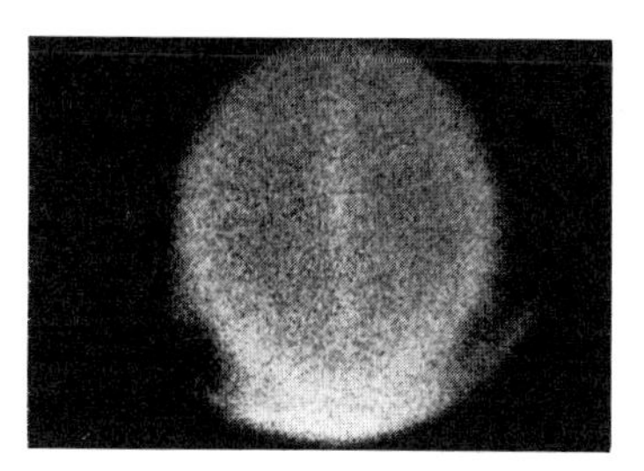

Fig. 8. Normal vertex brain scan. As the study was performed with ^{99m}Tc-DTPA, which does not accumulate in the salivary glands as does pertechnetate, atropine administration is not necessary. Note the extent to which the midline superior sagittal sinus is visualized.

the tumor, the intercellular spaces, the brain tumor cells, or a combination of these. Since the area of brain surrounding the tumor is usually relatively avascular, the relative absence of radioactivity in the uninvolved brain and even a slight increase of radioactivity in the tumor permits it to be demonstrated on the scan.

Localization of radionuclide in a tumor of the brain parenchyma apparently occurs gradually. Because of this, scanning should not be started immediately after injection because insufficient time will have elapsed for the radioactive material to enter the tumor.[2] Scans to detect a parenchymatous brain tumor are usually started 1 to 2 hours after technetium 99m injection.

In contradistinction to the tumors of the brain substance, the highly vascular meningiomas are probably visualized on scans because of the radioactive material in the vascular or endothelial spaces within these tumors. This is probably why meningiomas are clearly visible on a scan as early as 15 to 20 minutes after pertechnetate injection. This good imaging may persist for 2 to 3 hours, or as long as the blood level of radioactivity continues to be high.

The scan of a patient with a glioblastoma is seen in Figure 9. The study was performed 1 hour after intravenous injection of ^{99m}Tc-sodium pertechnetate. The space-occupying lesion is seen on both the anterior and right lateral scans. On the anterior scan the area of increased radioactivity is situated predominantly on the right side but crosses the midline. On the right lateral scan the tumor appears to be situated in the right frontal region and measures approximately 4 cm in diameter. Thus the scan identifies a rounded space-occuping lesion situated deeply in the right frontal region and extending slightly to the left of the midline.

Such descriptive identification of a brain lesion is important. In order to gain the greatest information concerning the size and location of the abnormality, visualization should be accomplished on an anterior or posterior scan *and* a lateral scan. The anterior and posterior scans may indicate the depth of

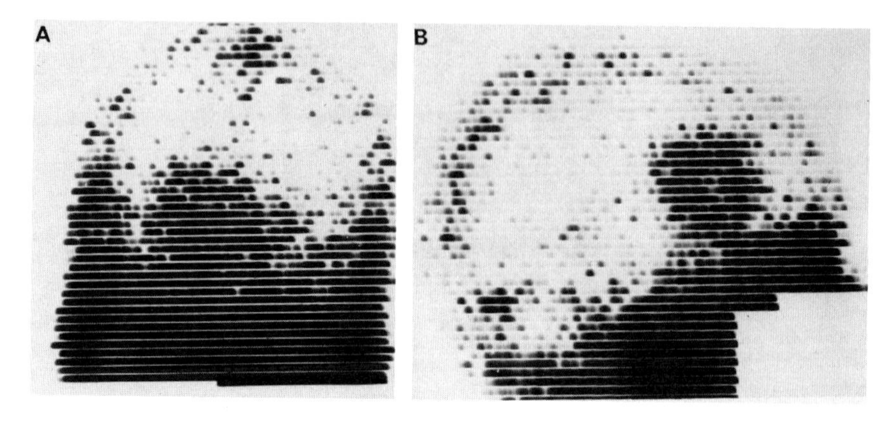

Fig. 9. Glioblastoma demonstrated 1 hour after ^{99m}Tc injection on anterior (A) and right lateral (B) scans.

the lesion and whether it crosses the midline, while a lateral scan helps identify its location. The size may be estimated in the two positions in which the abnormality is best defined.

However, demonstration of a brain tumor is not always so clear-cut. For example, in a patient hospitalized because of gradual onset of a left hemiparesis, initial studies performed 1 hour following ^{99m}Tc injection showed a normal anterior scan and a right lateral scan with minimally increased radioactivity in the right frontal region (Figs. 10A and B). Delayed scans performed 4 hours after pertechnetate injection showed a far different picture. The delayed anterior scan shows a somewhat rounded area of increased radioactivity on the right side (Figs. 10C and D). The right lateral scan now shows evidence of a fairly well defined space-occupying lesion in the right frontal region. This was found to be a glioblastoma at surgery.

Why could the abnormality be seen on the delayed scan, with no such information on the 1-hour study? It may be because there was gradual localization of the radionuclide in the region of the brain tumor. In this case it took approximately 4 hours for sufficient pertechnetate to accumulate in the glioblastoma so that it could be visualized on the scan.

This phenomenon of delayed brain scanning currently has wide application in the field of nuclear medicine.[3] It should be pointed out that a tumor that can be visualized only on a delayed scan at 3 to 4 hours is usually imaged with far less clarity than a tumor that can be seen well during the first hour after tracer administration. This is due in part to continued decay and excretion of the ^{99m}Tc. It may be that there is a different optimum scanning time for each type of brain tumor. However, although some tumors may not show scan visualization until 3 to 4 hours following radionuclide injection, other

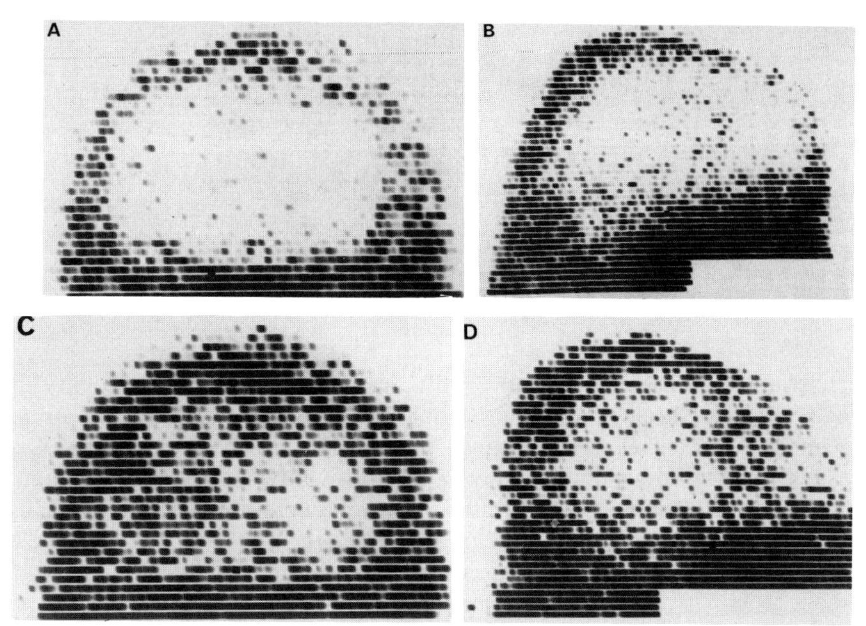

Fig. 10. Glioblastoma demonstrated on delayed scans only. A and B. Initial scans, performed 1 hour after pertechnetate injection. There is no discrete evidence of tumor in the anterior (A) or right lateral (B) positions. C and D. Scans taken 4 hours after injection. These define the area of tumor in the anterior (C) and right lateral (D) positions.

patients with tumors of the same type show good tumor visualization within the first hour. This is well seen in the two patients with glioblastomas just discussed.

For the most part, those supratentorial tumors involving the cerebral hemispheres may be imaged with clarity. These include gliomas (Fig. 11) of

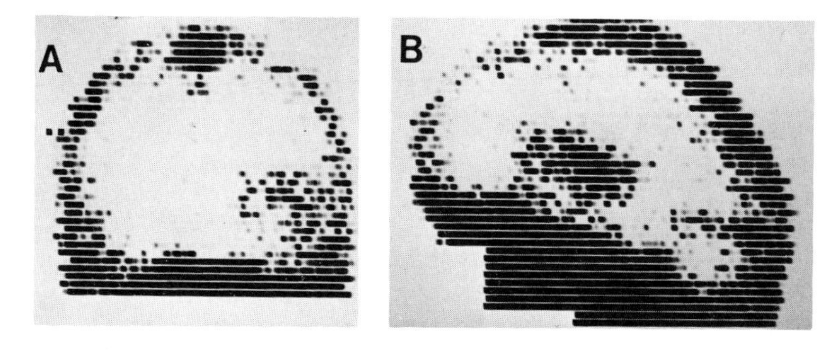

Fig. 11. Left frontotemporal glioblastoma in the anterior (A) and left lateral (B) positions. Imaging was performed 1 hour after ^{99m}Tc injection.

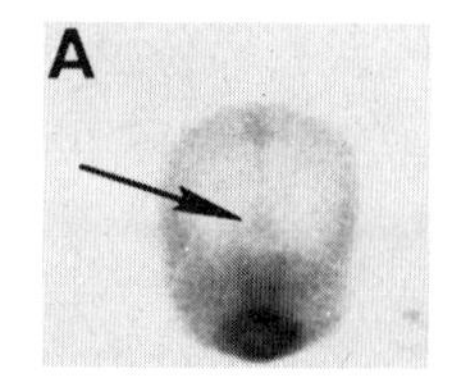

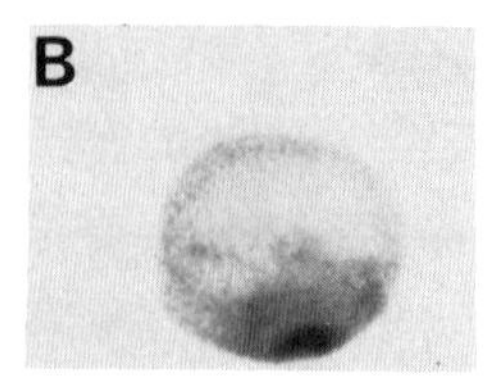

Fig. 12. Craniopharyngioma in the anterior (A) and right lateral (B) positions. Note the midline activity on the anterior scan.

all grades, tumors arising in the ventricles, and metastases. The ability to demonstrate a brain tumor with radionuclide imaging does not seem to be related to the degree of malignancy.

On the other hand, there may be great difficulty in demonstrating some supratentorial tumors closely related to the midline, such as those situated in the suprasellar and parasellar areas. Although craniopharyngiomas (Fig. 12) and chromophobe adenomas can usually be seen on scans, the imaging of these lesions is still relatively poor.

Metastases to the brain usually have the same imaging characteristics as primary tumors. The probability of metastatic lesions is likely when more than one space-occupying lesion is visualized (Fig. 13).

Meningiomas are probably visualized better and more consistently on scans than any of the other intracranial space-occupying lesions. As pointed out earlier they may be well seen as early as 15 minutes following tracer injection, and this good imaging may persist for 2 to 3 hours.

Radionuclide images of patients with convexity meningiomas are shown on Figure 14. Note the clarity with which such lesions are seen. On the left lateral scan illustrated in Figure 14B, the tumor is seen to be situated in the left frontoparietal region, superiorly, while the scan in the anterior position (Fig. 14A) shows the lesion to extend medially over the convexity. Imaging of another patient with a left convexity meningioma is seen in Figures 14C and

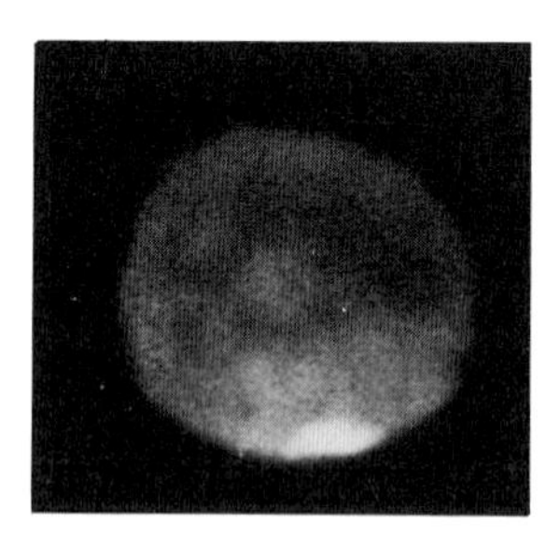

Fig. 13. Two space-occupying lesions representing metastases from carcinoma of the uterus.

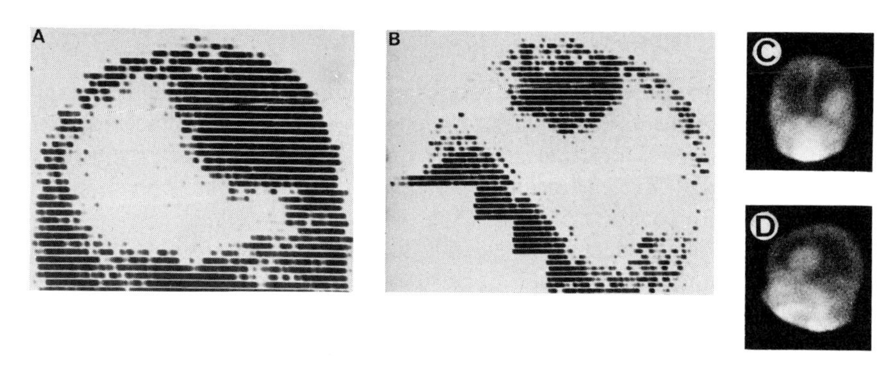

Fig. 14. A and B. Convexity meningioma in the anterior (A) and left lateral (B) positions. Scanning was started 15 minutes after pertechnetate injection. C and D. Left frontal convexity meningioma in the (C) anterior and (D) left lateral positions performed on a gamma camera 1 hour after ^{99m}Tc administration.

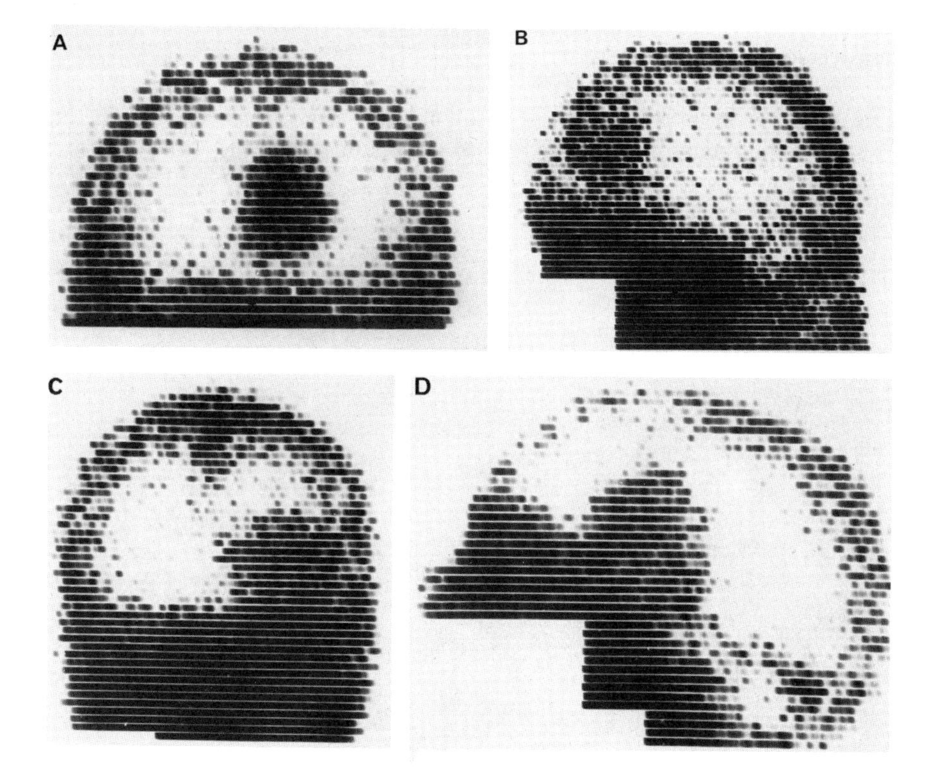

Fig. 15. A and B. Frontal meningioma seen in the anterior (A) and left lateral (B) positions. C and D. Sphenoid wing meningioma seen in the anterior (C) and left lateral (D) positions.

D. Scans demonstrating sphenoid wing and frontal meningiomas are seen in Figure 15.

Posterior Fossa Lesions. The posterior cranial fossa is visualized on a scan as the region inferior to the transverse sinuses and is best seen in the posterior and lateral positions. The tumors in this area most frequently demonstrated with nuclear medicine imaging are those of the cerebellopontine angle (e.g., acoustic neuromas) and those of the cerebellum.

It is important to determine whether the posterior fossa tumor is located in a relatively anterior or posterior position on the lateral scan; this is indicated by its relation to the projection of the transverse sinus and torcular Herophili. If the tumor appears to underlie the *anterior* portion of the transverse sinus, it is probably located in the region of the cerebellopontine angle (Fig. 16). Those lesions showing increased radioactivity in relation to the midportion of the transverse sinus region are usually so situated that they involve the vermis alone or both vermis and cerebellar hemisphere. Finally, areas of increased radioactivity closely related to the torcular Herophili image are usually situated in a cerebellar hemisphere as that structure extends posteriorly.

Of equal importance is the space-occupying lesion's location in the posterior position. Although there are notable exceptions, tumors of the cerebellopontine angle do not extend to the midline when they are small. On

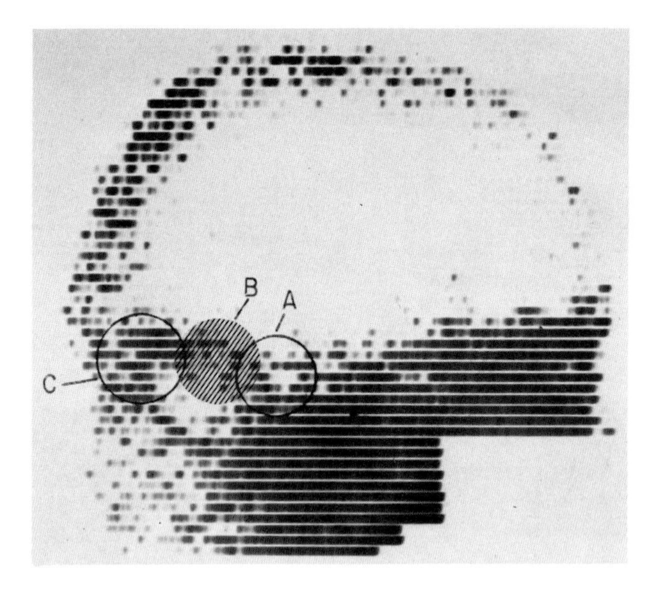

Fig. 16. Relation of posterior fossa lesions to their position on the lateral scan: region of cerebellopontine angle tumors (A); region of lesions involving a cerebellar hemisphere and the vermis (B); region of tumors involving only a cerebellar hemisphere (C).

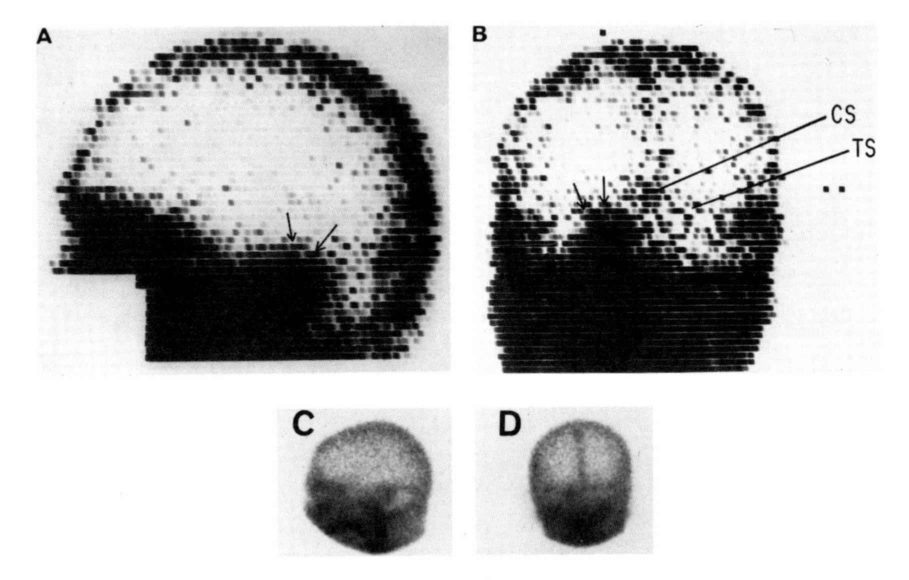

Fig. 17. Acoustic neuroma 2.5 hours after pertechnetate administration. A. Left lateral scan. The tumor (arrows) is seen in the relative anterior portion of the posterior fossa area. B. Posterior scan. Note the position of the confluence of the sinuses (CS) and the transverse sinuses (TS) in relation to the tumor (arrows). (From Baum et al: J Neurosurg 36:141, 1972.) C and D. Scintillation camera images of a left-sided acoustic neuroma in another patient seen in the left lateral (C) and posterior (D) positions.

the other hand, tumors of a cerebellar hemisphere are often closely related to the midline, as are of course those involving the vermis.

Scans of a patient with a left-sided acoustic neuroma are seen in Figure 17. Note the anterior location of the tumor on the left lateral scan. On the posterior scan the tumor lies in the posterior fossa region and is well to the left of the midline. Observe its relationship to the faintly outlined confluence and transverse sinuses.

A tumor involving the left cerebellar hemisphere and vermis is shown in Figure 18. This tumor was found to be an astrocytoma at surgery. The scan of another patient with an astrocytoma—in this instance confined to the right cerebellar hemisphere, posteriorly—is shown in Figure 19. The images of the tumor and torcular Herophili appear to be confluent.

Vascular-Type Lesions

The detection of vascular-type lesions, such as cerebral infarctions and subdural hematomas, is now a frequently employed application of brain scan-

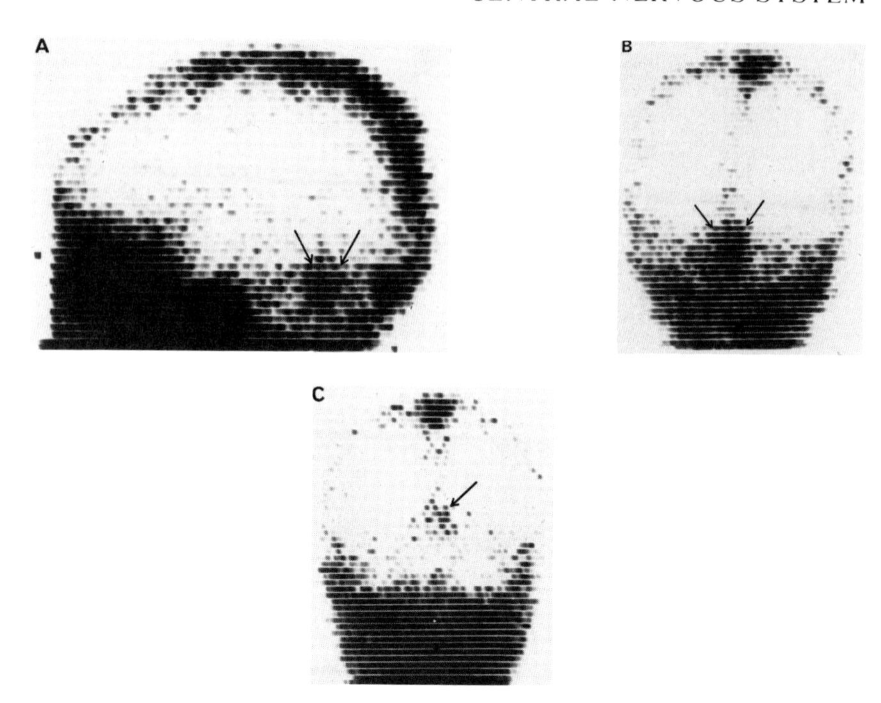

Fig. 18. A and B. Astrocytoma (arrows) involving the left cerebellar hemisphere and vermis in the left lateral (A) and posterior (B) positions. C. Scan performed 6 months after tumor removal. Note the position of the confluence (arrow).

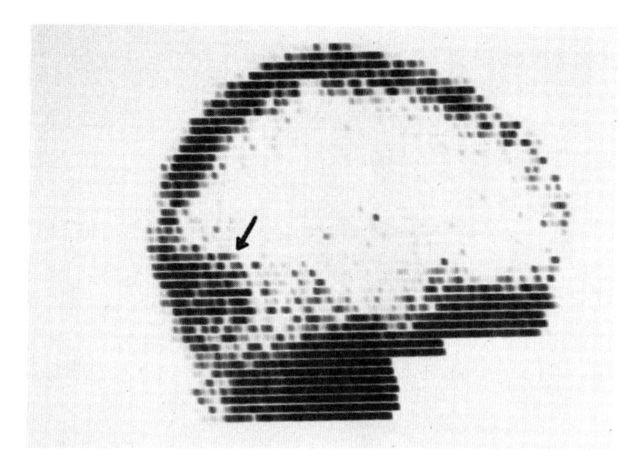

Fig. 19. Astrocytoma (arrow) of the right cerebellar hemisphere (not including the vermis) in the right lateral position.

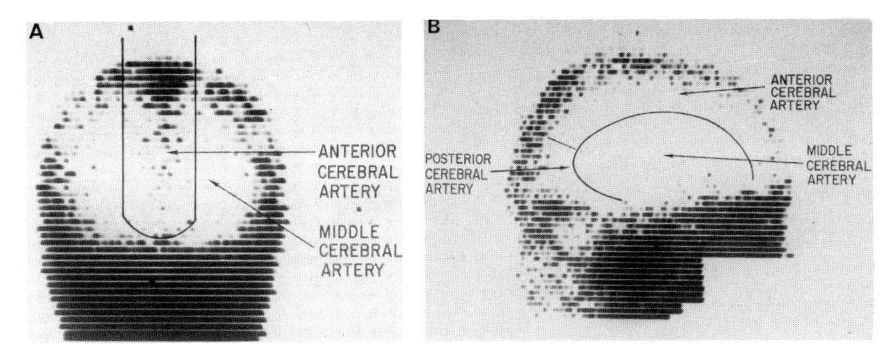

Fig. 20. Distributions of the major cerebral arteries as they are related to the anterior (A) and lateral (B) brain scans. (After DeLand.[4])

ning. The distributions of the major cerebral arteries as they are related to a brain scan are shown in Figure 20.[4] In addition to identifying the general area of involvement, knowledge of such distribution may help clarify the type of lesion involved. This is especially important in attempting to distinguish cerebral infarctions and subdural hematomas in some cases.

Cerebral Infarction. After an embolus is lodged in a cerebral artery or following formation of a thrombus, the area of brain distal to the occlusion is rendered ischemic. If a radioactive material such as ^{99m}Tc is injected immediately after the artery is occluded, no area of abnormally increased radioactivity is detected on the scan because no structure or structural alteration has yet developed to which the radionuclide may gain entry. However, over the following days the ischemic area becomes necrotic, and proliferating fibroblasts and capillaries surround it (Fig. 21). This zone becomes well delineated by 7 to 10 days after the initial insult; rarely it might be clearly seen as early as 3 to 4 days. The development of this new structure is probably associated with an alteration in the blood-brain barrier. If a radioactive material is administered at this time, the tracer localizes in the infarcted area because now a structure has formed to which the radionuclide may readily gain access.

The processes by which the tracer enters and accumulates are poorly understood. The radionuclide probably enters the proliferative zone surrounding the infarct and then gradually accumulates in the necrotic area, the latter likely occurring over a period of several hours.

An apt example is the case of a 56-year-old man with a cardiac arrhythmia who was hospitalized because of the sudden onset of a right hemiplegia. The brain scan 1 day after the onset of symptoms shows no distinct abnormality (Fig. 22). However, 2 weeks later the scan showed an

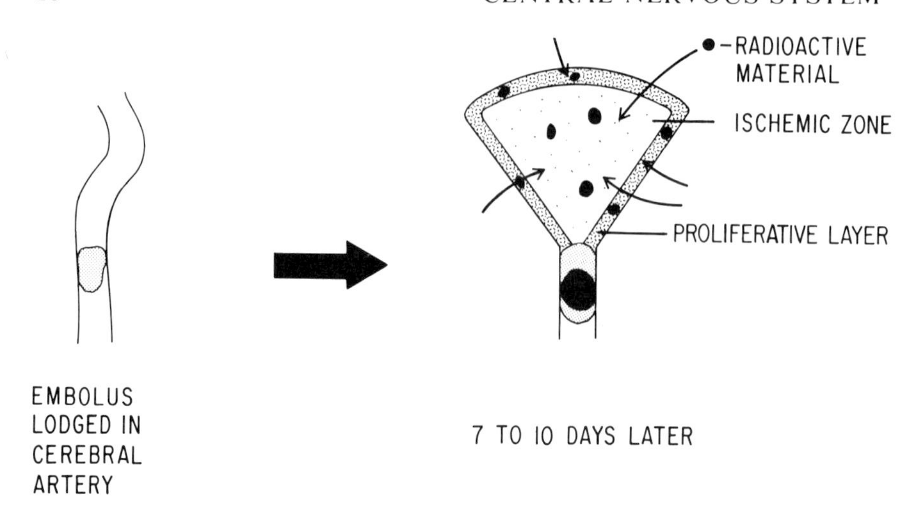

Fig. 21. Development of cerebral infarction with subsequent radionuclide entry. The radioactive material can enter the infarcted area only after a structural alteration has developed.

oblong area of increased radioactivity in the left frontotemporal region, in the distribution of the left middle cerebral artery. This sequence certainly suggests a cerebral infarction. At the time of the first scan a structural alteration was not yet present, but at the time of the second study it was sufficiently developed to allow the tracer to enter it.

Cerebral infarctions are usually best seen 2 to 4 weeks after onset and then usually show evidence of resolution. The necrotic area shrinks, and by 8 weeks after the initial episode the scan usually shows the infarct to be significantly reduced in size (Fig. 23). The structural and functional changes described apply to both ischemic and hemorrhagic infarcts.

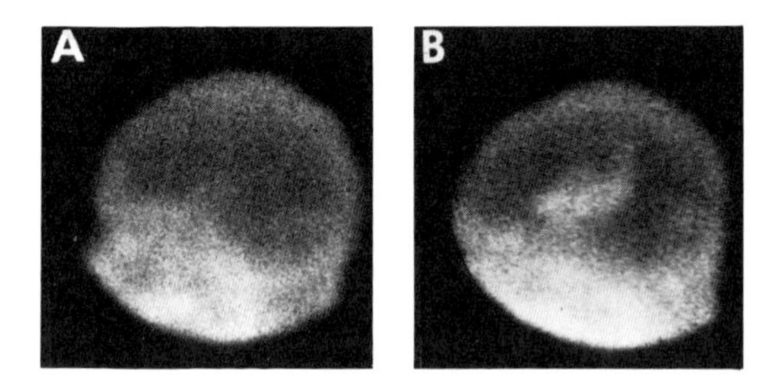

Fig. 22. Cerebral infarction. A. Left lateral scan 1 day after onset of right hemiplegia. B. Scan of same patient 2 weeks later showing evidence of infarction.

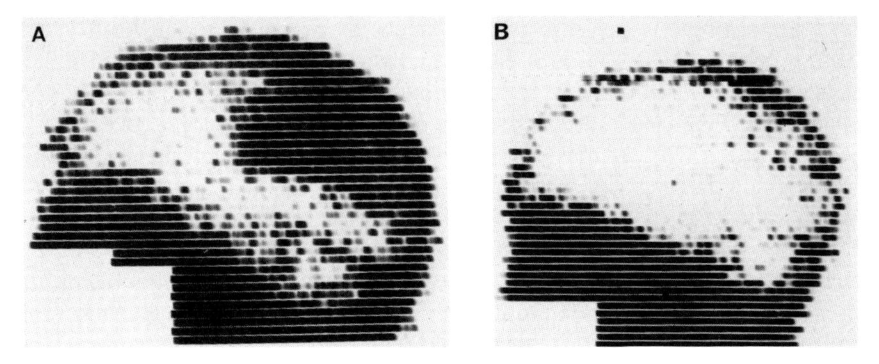

Fig. 23. Resolution of cerebral infarction. A. Three weeks after onset of right hemiparesis. B. Infarction is resolving 6 weeks later.

Subdural Hematoma. Following the shearing of cortical veins in the subdural space, usually from trauma, there is bleeding with subsequent formation of a clot. A highly vascular membrane forms around the clot, which in turn undergoes liquefaction and by its osmotic activity draws fluid into the confines of the membrane (Fig. 24). This subdural membrane takes about 7 to 10 days to form. If ^{99m}Tc-sodium pertechnetate is administered after this vascular membrane has formed, radioactive material accumulates in the membrane and in the subdural fluid within the membrane. This appears as an area of increased radioactivity on the scan. Attempts at scanning prior to formation of the membrane, when only the clot is present, are unrewarding because no structure is yet formed to alter the blood-brain barrier and in which the tracer could accumulate. Thus the acute subdural hematoma is not detected by static radionuclide imaging.

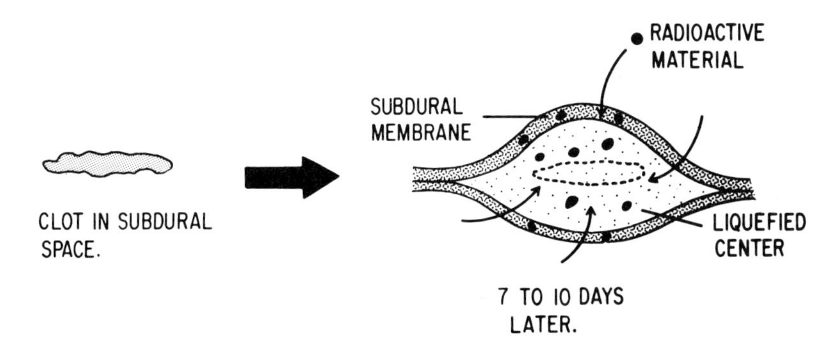

Fig. 24. Development of subdural hematoma and subsequent radionuclide entry. Accumulation of radionuclide must await formation of the subdural membrane.

An example is the case of a 70-year-old woman who had been in an automobile accident 3 months earlier and was hospitalized because of right-sided weakness and occasional confusion of 2 weeks' duration. The anterior scan showed increased radioactivity on the left side, peripherally, over a very wide area (Fig. 25). Increased radioactivity on the left side, peripherally, was also seen on the posterior scan. On the left lateral scan increased radioactivity was barely detected over the left frontal and parietal regions. Subsequently a large left-sided subdural hematoma was evacuated.

Subdural hematomas usually are seen on scans as peripheral areas of increased radioactivity, often crescent-shaped.[5] The lesion is imaged very well in the anterior and posterior positions because here the abnormality is seen in its entire depth. However, on the lateral scan the increased radioactivity is of low intensity because a relatively thin membranous structure is visualized in this position. The optimum time for scanning is approximately 1 to 2 hours after ^{99m}Tc injection when there is sufficient radionuclide accumulation.

Unlike cerebral infarctions, the subdural hematoma is unrelated to the

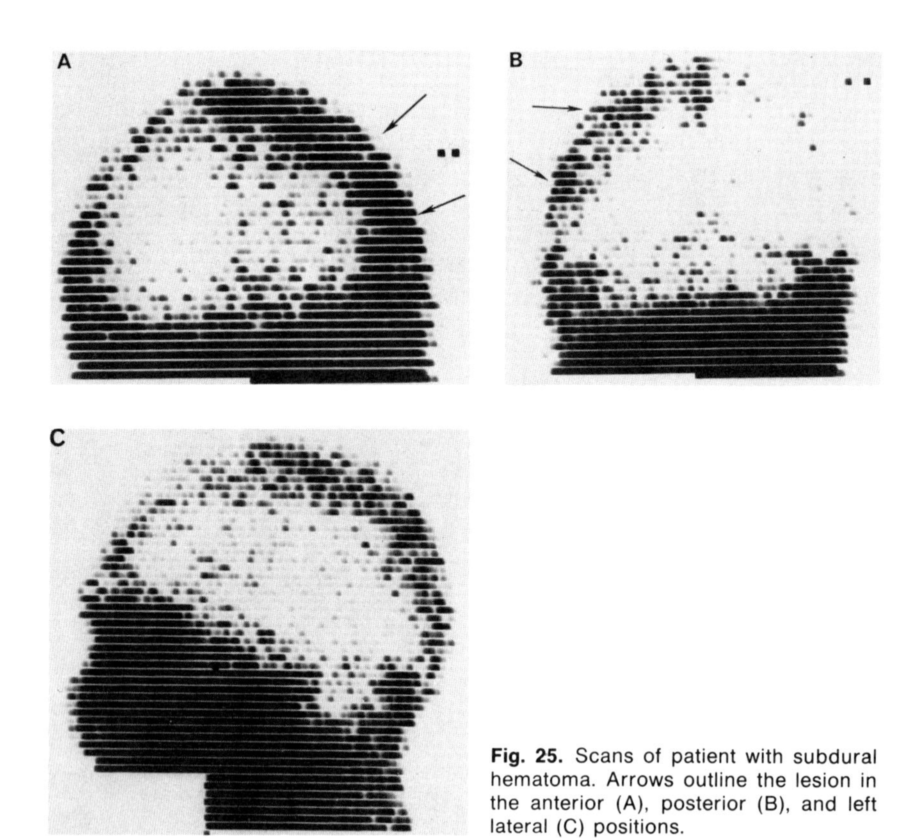

Fig. 25. Scans of patient with subdural hematoma. Arrows outline the lesion in the anterior (A), posterior (B), and left lateral (C) positions.

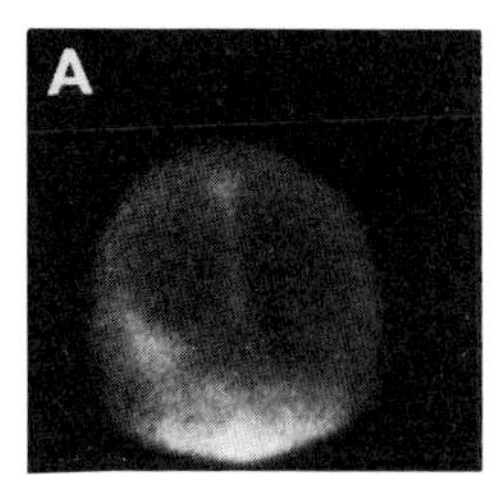

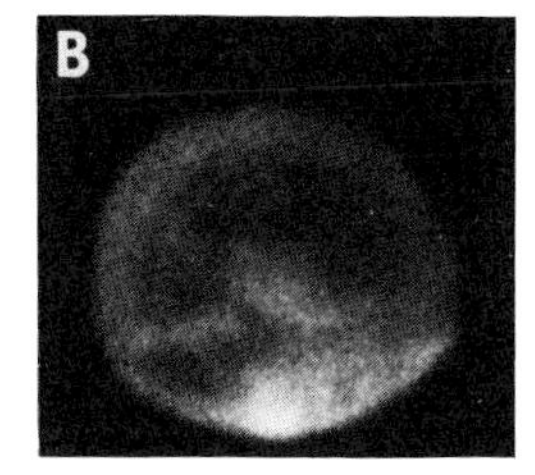

Fig. 26. Arteriovenous malformation demonstrated 30 minutes after ^{99m}Tc injection in the anterior (A) and right lateral (B) positions. The linear area of increased radioactivity just posterior to the lesion, as seen in the right lateral position, represents a large draining vein.

distribution of the cerebral arteries. In the subdural hematoma just described the area involved on the anterior scan overlies the distribution of both the anterior and middle cerebral arteries. It would be virtually impossible for an embolus to involve these two major vessels at one time, and this wide area of increased radioactivity helps rule out the possibility of cerebral infarction.

Another common vascular-type lesion is the intracerebral hemorrhage. Like the cerebral infarction and subdural hematoma, a sheath forms around the hemorrhage about 7 to 10 days after it appears, and at this time radioactive material can accumulate in the surrounding vascular layer and in the hemorrhagic area itself. The scan shows a localized, rounded area of increased radioactivity.

Countless other vascular-type lesions may be seen on scans, and in each case there is an alteration in the blood-brain barrier allowing a tracer to enter and accumulate in the involved area. Common examples are cerebral contusion, epidural hematoma, postcraniotomy changes, and cortical atrophy.

A special type of vascular lesion is the arteriovenous malformation. Because of its nature, it is best visualized immediately after radionuclide injection when the level of radioactivity is at its height. It is seen on the scan as a localized area of increased radioactivity and may resemble a space-occupying lesion (Fig. 26).

Nonneoplastic Space-Occupying Lesions

The principal nonneoplastic space-occupying lesions demonstrated on scans include brain abscess and granuloma (e.g., tuberculomas). Most often on the scan they are indistinguishable from a tumor.

Scans of a patient with an abscess in the left frontal lobc are seen in Figure 27. The abscess is probably visualized initially as radioactive material enters the vascular network associated with the abscess capsule. Delayed

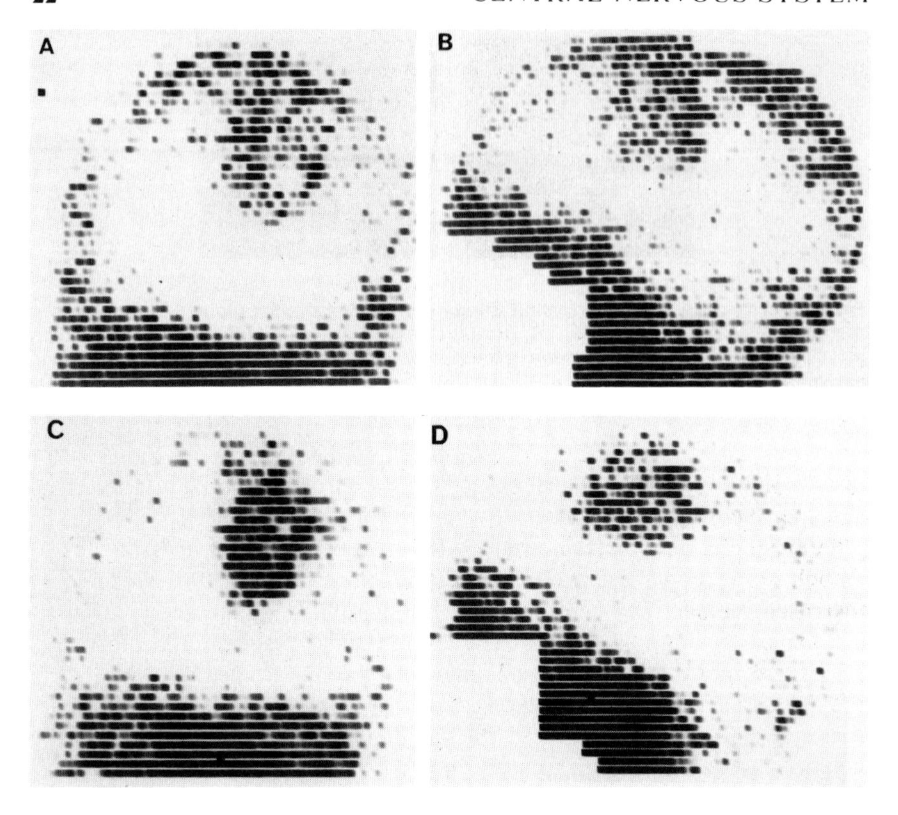

Fig. 27. Scans of patient with posterior frontal abscess. A and B. One hour after pertechnetate injection. C and D. Four hours after radionuclide was given.

scans may show continued accumulation of radionuclide in the central abscess cavity.

Tuberculomas and sarcoid granulomas are easily visualized on brain scans. These granulomatous lesions are often multiple and may resemble the scan appearance of metastases.

Clinical Considerations

Brain Tumors. The initial approach to the patient suspected of having an intracranial space-occupying lesion is centered around diagnostic procedures that can detect such a lesion with a minimum of discomfort and morbidity to the patient. The brain scan is included in this category along with the skull x-ray and electroencephalogram.

If a tumor is visualized on the 1-hour scan and a primary lesion is suspected, further roentgenologic studies are usually undertaken to help define the abnormality. This is most frequently accomplished with cerebral angiography. The precise anatomic information in regard to location, displacement of blood vessels, relation to adjacent structures, and probable tumor type can usually be supplied only by roentgenologic methods, the brain scan being no substitute for them. A nonneoplastic lesion on the scan may resemble a brain tumor, and often only by cerebral angiography can the distinction be made with accuracy.

Such specific roentgenologic identification is usually not necessary in those patients with known primary neoplasms who are thought to have intracranial metastases. The single or multiple space-occupying lesions seen on brain scans in these cases are usually considered to be metastatic.

One of the most complex situations arises with a patient suspected of having a brain tumor but whose scan at 1 hour does not demonstrate such a lesion. How such a case should be handled in regard to brain scanning is still open to question. It seems reasonable that if the possibility of a brain tumor is considered the search for it should be pursued, which may be accomplished in the following manner.

If the 1-hour scan is negative in the brain tumor suspect, scans should be performed in the suspected lateral and posterior or anterior positions at 2 to 3 hours following radionuclide administration. If specific diagnostic localization of the possible tumor can not be determined, the scan should be performed in all four conventional positions. If an abnormality is detected on the delayed study, further scanning is usually not necessary. However, failure to detect a space-occupying lesion at 3 hours does not indicate that a tumor definitely is not present. It means only that scanning can not demonstrate the lesion, and if one is thought to exist then roentgenologic procedures should be performed.

There are some intracranial neoplasms that often defy scan detection at the present time, no matter how the scanning times and techniques are altered. These include many of the midline tumors. It may be their relatively deep position and small size that makes imaging difficult. Of course, those lesions less than 2 cm in diameter will probably remain undetected.

Vascular-Type Lesions. The performance of brain scans in patients with strokes may be helpful in detecting surgically correctable vascular-type lesions (e.g., subdural hematoma). If the scan shows an area of increased radioactivity, especially of the peripheral crescent pattern, immediately after the onset of symptoms, the possibility of a chronic subdural hematoma should be considered. Something must have occurred in the past (usually at least 7 to 10 days previously) that allowed a vascular membrane to form, providing entry for the radionuclide. If, on the other hand, the scan does not show evidence of increased radioactivity until a week or two after the onset of symptoms, then a lesion such as a cerebral infarction or intracerebral hemorrhage is more likely. In any case, if subdural hematoma is suspected it is essential that confirmation be made by cerebral angiography. Too many other abnormalities may resemble the scan appearance of a subdural hematoma to have complete confidence in the scan as a diagnostic technique of choice. It is often difficult or impossible to distinguish between neoplastic and nonneoplastic lesions on scans. A cerebral infarction may look very much like a brain tumor; but the brain tumor does not become smaller after a period of time, whereas the infarction often shows signs of resolution. Interpretation should be tempered and altered by clinical suspicion.

RADIONUCLIDE CISTERNOGRAPHY

A more recent innovation in radionuclide studies of the central nervous system has been the development of radionuclide cisternography.[6] In this procedure a radioactive material is injected into the cerebrospinal fluid (CSF) system, usually by introduction into the lumbar subarachnoid space or less commonly through direct cisternal puncture.

For the most part, the site of CSF formation is the choroid plexus located in the lateral and third ventricles. This fluid bathes the ventricular system and the lumbar and intracranial subarachnoid spaces. It is absorbed principally through the arachnoid villi as these structures protrude into the venous sinuses of the dura mater.

Among the radionuclides commonly employed in cisternography are labeled proteins, principally ^{131}I-human serum albumin (HSA) and less commonly ^{99m}Tc-HSA. The latter is not frequently used because its relatively short half-life makes it difficult to perform imaging 24 to 48 hours following tracer administration. Since intrathecal administration of a large amount of HSA may result in a chemical or aseptic meningitis, it is essential to use HSA of high specific activity so the patient will receive no more than 4 mg of protein.[7] The usual adult dose of ^{131}I-HSA is 100 μCi and of ^{99m}Tc-HSA is 1 to 2 mCi. Recently, DTPA (diethylenetriaminepentaacetic acid), a chelating agent, tagged with radioactive indium has been used for cisternography. Although both ^{113m}In, with a half-life of 99 minutes, and ^{111}In, with a half-life of 2.8 days, have been employed, the ^{111}In-DTPA, in a dose of about 500 μCi, is preferred because of its longer half-life.

Another chelated complex that is being used with increasing frequency is ytterbium 169 DTPA. The gamma emissions at 177 keV and 198 keV may be utilized for imaging. Although the physical half-life of ^{169}Yb is relatively long (32 days), the effective half-life of ^{169}Yb-DTPA in the intracranial sub-

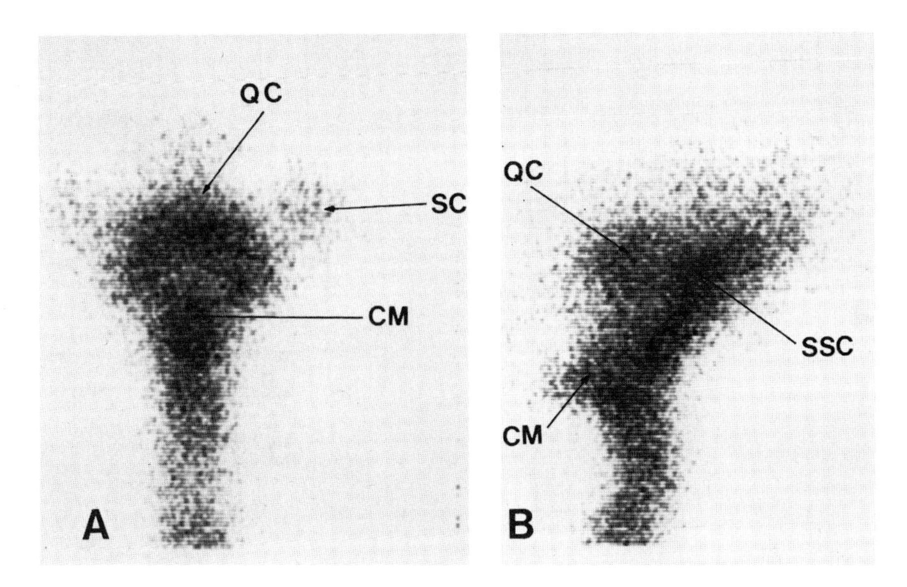

Fig. 28. Radionuclide cisternogram performed 2 hours following intrathecal administration of ^{169}Yb-DTPA in the anterior (A) and Rl right lateral (B) positions. CM-cisterna magna; QC-quadrigeminal cistern; SC-Sylvian cistern; SSC-suprasellar cistern. (Courtesy of Dr. Norman R. Vincent, St. Vincent's Hospital, Bridgeport, Conn.)

arachnoid space in the healthy subject is thought to be much shorter. The usual intrathecal dose in the adult is 1 mCi.

After instillation of the radioactive material into the lumbar subarachnoid space, images of the cranial region in the anterior and one lateral position are normally obtained at 2, 6, and 24 hours.

The 2-hour scan usually shows evidence of radioactive material in the cisterna magna and other basal cisterns (Fig. 28).[8] By 6 hours the Sylvian cisterns are well outlined (Fig. 29). The radioactive material then appears to travel through the cortical subarachnoid pathways as it finally enters the arachnoid villi to be absorbed. The 24-hour study thus shows evidence of radioactivity in the parasagittal region and over the convexities (Fig. 30); such findings may sometimes appear as early as 6 hours after the tracer is given. It is important to recognize that at no time are the lateral ventricles visualized on the normal radionuclide cisternogram.

Normal-Pressure Communicating Hydrocephalus

The initial impetus to the development and sophistication of cisternography was the recognition of normal-pressure communicating

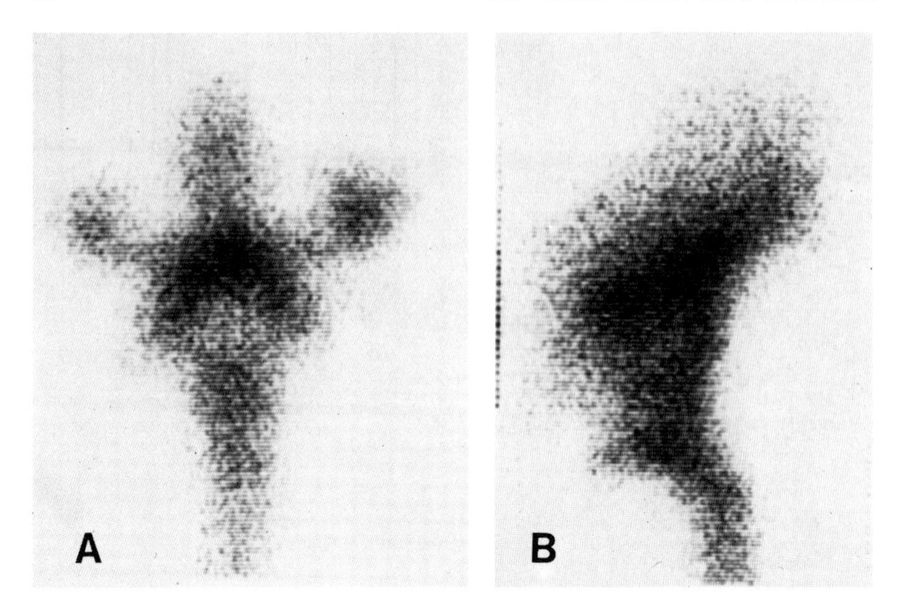

Fig. 29. Radionuclide cisternogram in the (A) anterior and (B) right lateral positions 6 hours following the administration of ^{169}Yb-DTPA. The Sylvian cisterns are now prominent. (Courtesy of Dr. Norman R. Vincent, St. Vincent's Hospital, Bridgeport, Conn.).

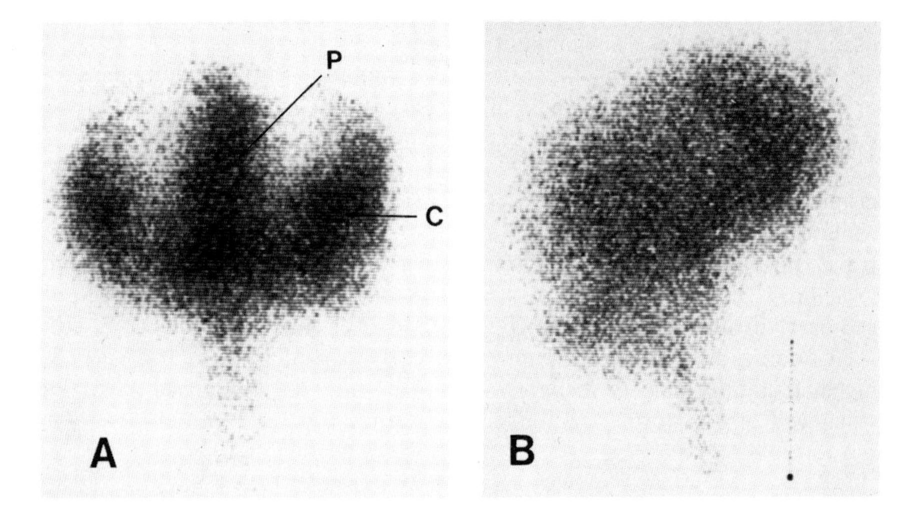

Fig. 30. Radionuclide cisternogram in the (A) anterior and (B) right lateral positions 24 hours following ^{169}Yb-DTPA administration. The radioactive material is now seen to be present over the convexities (C) and the parasagittal region (P). (Courtesy of Dr. Norman R. Vincent, St. Vincent's Hospital, Bridgeport, Conn.).

hydrocephalus.[9] In this syndrome the usual flow of CSF is obstructed somewhere in the intracranial subarachnoid space, resulting in a reversal of CSF flow and dilatation of the lateral ventricles. There is free communication between the ventricular system and the subarachnoid pathways (Fig. 31) and no elevation of CSF pressure—hence the term normal-pressure communicating hydrocephalus.

The entity is characterized clinically by dementia, gait disturbances, and fecal and urinary incontinence. However, decompression of the ventricles or a shunting procedure in many cases results in dramatic clinical improvement of the patient. Although there are numerous causes for the obstructive phenomena that may be responsible for normal-pressure hydrocephalus, they most commonly result from subarachnoid hemorrhage or meningoencephalitis.

There are striking cisternographic findings in the patient with normal-pressure communicating hydrocephalus.[10] Imaging performed as early as 2 to 4 hours following radionuclide administration shows evidence of radioactivity in the lateral ventricles, and this good visualization may persist for 24 to 48 hours (Fig. 32). There is delayed clearance of the radioactive material as evidenced by only minor visualization of radioactivity over the cerebral convexities as late as 24 to 48 hours.

The reasons for these altered cisternographic findings in patients with normal-pressure communicating hydrocephalus are still not completely understood. Because of the obstruction in the subarachnoid space there is probably a reversal of CSF flow, and instead of the primary avenue of CSF absorption occurring at the arachnoid villus there is also major reabsorption in the lateral ventricles through the lining ependymal cells.[11] This may serve to explain the accumulation of the tagged tracer in the lateral ventricles (Fig. 31).

Clinical Considerations. Although the described cisternographic findings are usually present in a patient with normal-pressure communicating hydrocephalus, they are by no means in themselves diagnostic. This study is but one of several procedures used to determine the presence or absence of this entity in which the age of onset of hydrocephalus is relatively late.

As stated earlier, the cardinal clinical manifestations are dementia, incontinence of urine and feces, and abnormalities of gait. The pneumoencephalogram shows dilatation of the ventricles, especially in the region of the anterior horn, and no air is seen over the sulci and gyri of the cerebral cortex.[11] Shunting results in dramatic clinical improvement.

Other causes of late-onset hydrocephalus must be considered, such as cortical atrophy. In the latter situation the pneumoencephalogram usually shows air over the sulci and gyri of the cerebral cortex in addition to dilated ventricles. Radionuclide cisternograms in these patients at 4 hours usually show evidence of radioactive material in the lateral ventricles. These findings

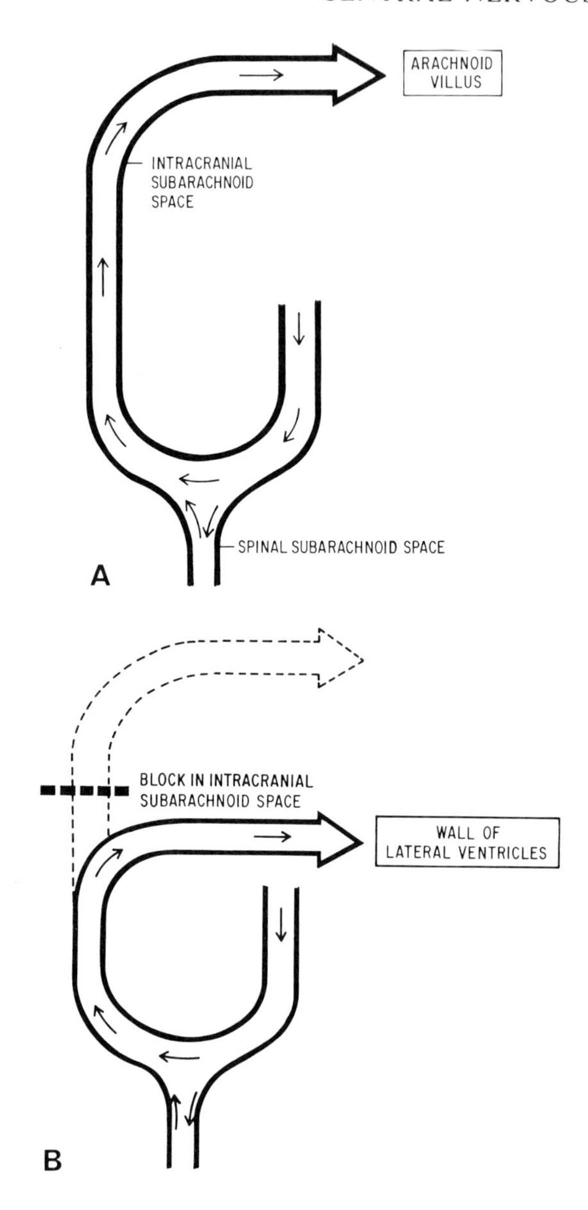

Fig. 31. Diagramatic representation of (A) normal cerebrospinal fluid flow and (B) the development of normal pressure, communicating hydrocephalus. Normally the CSF that is produced by the choroid plexus formation in the lateral ventricles becomes absorbed in the arachnoid villi. In normal pressure hydrocephalus obstruction in the intracranial subarachnoid space results in reversal of flow and absorption of CSF in the walls of the lateral ventricles.

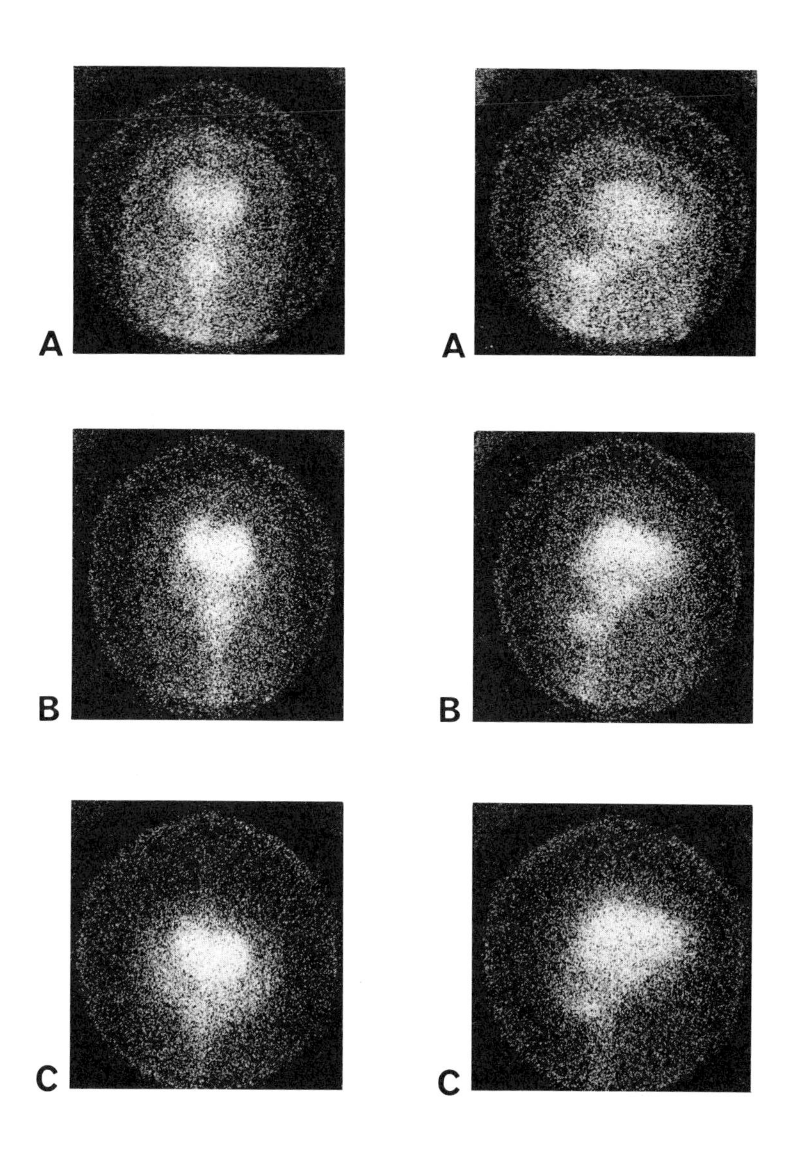

Fig. 32. Cisternogram of patient with normal pressure hydrocephalus showing evidence of radioactivity in the lateral ventricles in the anterior and right lateral positions using ^{169}Yb-DTPA. The lateral ventricles are well seen as early as (A) 2 hours following tracer instillation. There is continued accumulation of tracer in the lateral ventricles as seen in the study (B) performed at 6 hours. By 24 hours (C), there is no evidence of clearance of radionuclide and there is persistence of radioactivity in the lateral ventricles (Courtesy of Dr. Norman R. Vincent, St. Vincent's Hospital, Bridgeport, Conn.).

often do not persist and by 24 to 48 hours radioactivity is visualized over the cerebral convexities and in the parasagittal area, corresponding to the clearance although delayed, of the tracer.

If hydrocephalus occurs secondary to ventricular obstruction, radioactive material is unable to enter the lateral ventricles and the cisternographic findings are similar to those seen in the normal subject.

Other Applications of Cisternography

Any structural alteration related to the ventricles or intracranial subarachnoid space has the potential to be visualized on a radionuclide cisternogram. Entities such as porencephalic and Dandy-Walker cysts have been demonstrated with clarity. It is still too early to determine to what degree cisternography will be used in the future for identifying these and other related abnormalities.

Other studies associated with intrathecal administration of radionuclides have been employed in the investigation of CSF rhinorrhea and otorrhea. The site of a CSF leak may be imaged, and such a procedure may be easily used for screening when such entities are considered. Alternatively, radioactivity counts may be obtained from an absorbent material placed in the orifice in question to determine whether CSF is indeed leaking.

REFERENCES

1. Bakay L: Basic aspects of brain tumor localization by radioactive substances: a review of current concepts. J Neurosurg 27:239, 1967
2. Ramsey RG, Quinn JL III: Comparison of accuracy between initial and delayed ^{99m}Tc-pertechnetate brain scans. J Nucl Med 13:131, 1972
3. Gates GF, Dore EK, Taplin GV: Interval brain scanning with sodium pertechnetate Tc 99m for tumor detectability. JAMA 215:85, 1971
4. DeLand FH: Scanning in cerebral vascular disease. Semin Nucl Med 1:31, 1971
5. Heiser WJ, Quinn JL III, Mollihan WJ: The crescent pattern of increased radioactivity in brain scanning. Radiology 87:483, 1966
6. DiChiro G: New radiographic and isotope procedures in neurological diagnosis. JAMA 188:524, 1964
7. DiChiro G, Ashburn WL, Briner WH: Technetium Tc-99m serum albumin for cisternography. Arch Neurol 19:218, 1968
8. James AE Jr, DeLand FH, Hodges FJ III, et al: Cerebrospinal fluid (CSF) scanning cisternography. Am J Roentgenol Radium Ther Nucl Med 110:74, 1970

9. Adams RD, Fisher CM, Hakim S, et al: Symptomatic occult hydrocephalus with "normal" cerebrospinal-fluid pressure: a treatable syndrome. N Engl J Med 273:117, 1965
10. Bannister R, Guildford E, Kocen R: Isotope encephalography in diagnosis of dementia due to communicating hydrocephalus. Lancet 2:1014, 1967
11. Harbert JC: Radionuclide cisternography. Semin Nucl Med 1:90, 1971

ADDITIONAL READING

Bakay L, Klein DM (eds): Brain Tumor Scanning with Radioisotopes. Springfield, Charles C. Thomas, 1969

DeLand FH, Wagner HN Jr: Brain. In: Atlas of Nuclear Medicine, Vol I. Philadelphia, Saunders, 1969

Gilson AJ, Smoak WM III (eds): Central Nervous System Investigation with Radionuclides. Second Annual Nuclear Medicine Seminar. Springfield, Charles C. Thomas, 1971

Handa J: Dynamic Aspects of Brain Scanning. Baltimore, University Park Press, 1972

Potchen EJ, McCready VR (eds): Progress in nuclear medicine: Neuro-Nuclear Medicine. Baltimore, University Park Press, 1972

Schall GL, Quinn JL III: Brain scanning. In Blahd WH (ed): Nuclear Medicine, 2nd ed. New York, McGraw Hill, 1971, pp 236-294

Wagner HN Jr, Holmes RA: The nervous system. In Wagner HN Jr (ed): Principles of Nuclear Medicine. Philadelphia, Saunders, 1968, pp 655-695

Zeidler U, Kottke S, Hundeshagen H: Hirnszintigraphie. Berlin, Springer-Verlag, 1972 (in German)

chapter 2

LIVER AND SPLEEN

LIVER IMAGING

Liver imaging has provided a simple means for visualization of hepatic size, configuration, and pathology to a degree that was hitherto unknown through plain roentgenograms. Although liver size and shape may be estimated with a fair degree of accuracy in many cases from x-rays of the abdomen, information does not approach that obtained with radionuclide imaging.

Radionuclide scanning of the liver may be performed after a radioactive material enters either (1) the endothelial lining Kupffer cells or (2) the polygonal parenchymal cells. If a patient receives an intravenous injection of foreign particles that can circulate freely, such particulate matter is phagocytized in the reticuloendothelial system (RES). The principal site of phagocytosis is in the Kupffer cells of the liver, but this process also occurs in the macrophages of the spleen and in the bone marrow. To circulate freely, the foreign particles must not be of a diameter greater than that of the smallest capillaries. If a radioactive colloid is injected intravenously it also becomes engulfed by the phagocytic Kupffer cells. Currently most hepatic scanning involves the use of radioactive particles in colloidal form.

With the other technique a radionuclide such as ^{131}I-rose bengal enters the parenchymal cells and subsequently is excreted into the biliary system.

Radioactive Colloids

The earliest of the radiocolloids successfully employed in liver imaging on a large scale was ^{198}Au colloidal gold (half-life 2.7 days, principal gamma emission 411 keV). Because of the relatively high radiation exposure associated with its use, it was common to administer a dose of only 150 to 300 μCi. However, an advantage of this material was that because of its long half-life scanning could be performed as late as 48 hours following radionuclide injection.

Technetium ^{99m} sulfur colloid is now far and away the most commonly used radiocolloid for liver imaging. The particle size is less than 0.1μ in diameter. The usual adult dose is 1.0 to 2.0 mCi. In the normal subject there is almost maximum incorporation of this radionuclide in the Kupffer cells about 5 minutes or sooner following intravenous injection. Unless stated otherwise, all reference to hepatic imaging with radiocolloids in this section involves the use of ^{99m}Tc-sulfur colloid.

Shortly after the ^{99m}Tc-sulfur colloid is injected intravenously, the Kupffer cells become loaded with the radioactive material. As the capacity of the liver to take up radionuclide is exceeded, the remaining radiocolloid becomes phagocytized elsewhere in the RES.

In the healthy person such uptake occurs principally in the spleen, with much less phagocytosis seen in the bone marrow. However, in those disease states in which there is reduced phagocytic activity in the liver or both liver and spleen, the rejected ^{99m}Tc-sulfur colloid is taken up in large quantities in the RES cells of the bone marrow.

The fate of a radioactive colloid once it becomes engulfed in the phagocytic Kupffer cells or other cells of the RES is still not entirely clear. The colloidal particles probably remain intracellular until the cells are sloughed in the process of renewal. However, some of the particles may themselves be extruded. There is no evidence of any harmful effects from the presence in the RES of these materials that have undergone radioactive decay.

Normal Liver Imaging

Since the advent of hepatic imaging with ^{99m}Tc-sulfur colloid the liver is rarely visualized alone, and the study should properly be called a liver-spleen scan. Imaging is generally performed in the anterior, right lateral, and posterior positions.

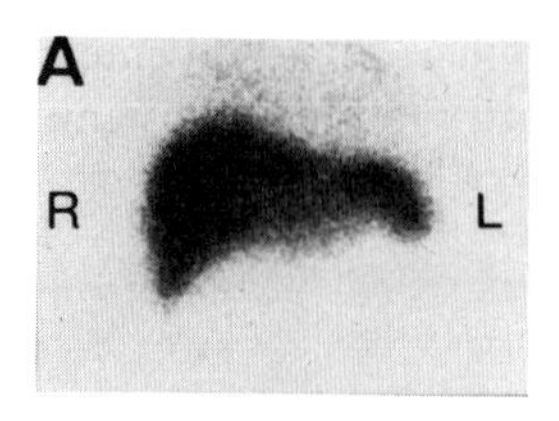

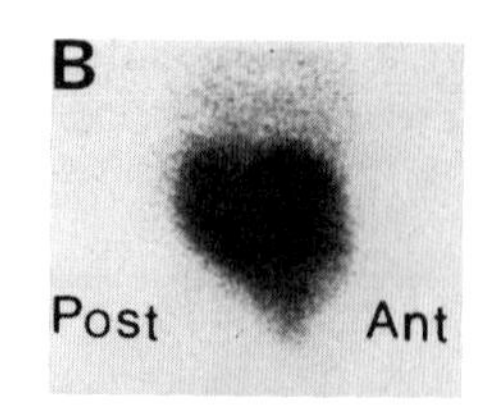

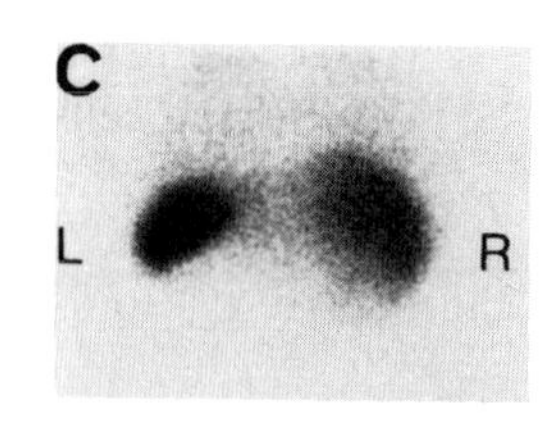

Fig. 1. Normal liver images obtained on a scintillation camera in the anterior (A), right lateral (B), and posterior (C), positions.

The configuration of the normal liver (Figs. 1 and 2) varies widely among patients. However, the type most commonly encountered is the organ with a larger right and a smaller left lobe. There are a number of characteristics associated with the normal liver-spleen scan:

1. Where the mass of the organ is greatest, the radioactivity is also greatest. Thus the right lobe appears to contain more radioactivity than the left. Also, the greatest mass and thus the greatest amount of radioactivity appears at the center of each lobe. Radioactivity is relatively reduced toward the periphery, as the organ thins.
2. The subdiaphragmatic portion of the right lobe is usually nicely rounded as it fits snugly underneath the right hemidiaphragm. An alteration in this curved appearance usually signifies an abnormality (e.g., flattening secondary to pulmonary emphysema, abscess).
3. A cardiac impression may be seen closely related to the medial portion of the right lobe and adjacent liver. This impression may vary with the size of the heart (Fig. 2B).
4. The separation or cleft between the smaller left and larger right lobe is often seen on the scan as an area of reduced radioactivity. This may mimic the appearance of a space-occupying lesion. The relatively reduced mass in the lower half of the left

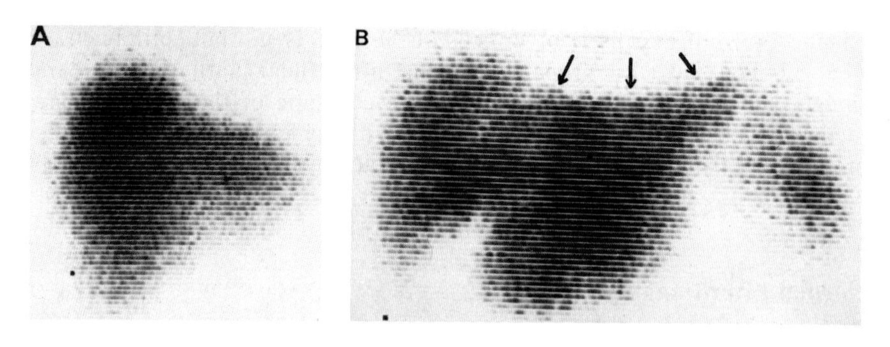

Fig. 2. A. Normal rectilinear liver scan performed in the anterior position. B. Anterior liver scan showing accentuation of the cardiac impression (arrows).

lobe may also show reduced radioactivity, which can mistakenly be associated with a space-occupying lesion.

5. The inferior portion of the right lobe that is not girded by the rigid thoracic cage frequently shows lateral extension on the anterior scan, and because of this the region superior to it may be mistaken for a space-occupying lesion. This is especially true in obese patients.
6. The liver usually has a somewhat triangular shape on the right lateral scan. Unless there is unusual anterior or posterior extension, delineation of the hepatic lobes in this position cannot easily be made.
7. Because of its relative anterior location and small mass, there is often only poor or nonvisualization of the left lobe on the posterior scan.
8. The spleen is easily visualized in all scanning positions. Note its characteristic posterior location on the right lateral scan. The subject of spleen scanning is treated more fully later in the chapter.

Size and Position. It is important to determine the size of the liver and its relation to the right costal margin. The latter is of inestimable value to the clinician. If a rectilinear scanner is used, the xiphoid and lowest rib are marked on the scan to identify roughly the right costal margin. The dimensions of the organ may be determined directly from the scan. It is common to measure the liver in its greatest longitudinal dimension, which normally does not exceed 20 cm. The left lobe usually extends into the epigastrium.

When using a scintillation camera a different method for marking and measuring is employed. A lead marker is placed over the lowest rib to identify the right costal margin (Fig. 3). For measurement, a lead ruler consisting of 2-cm lead markers, alternating with 2-cm blank areas is placed over the patient's abdomen, and the dimensions are thus calculated.

Variants of Normal. It is well known that the shape of the liver varies widely, and countless hepatic configurations have been described. In some

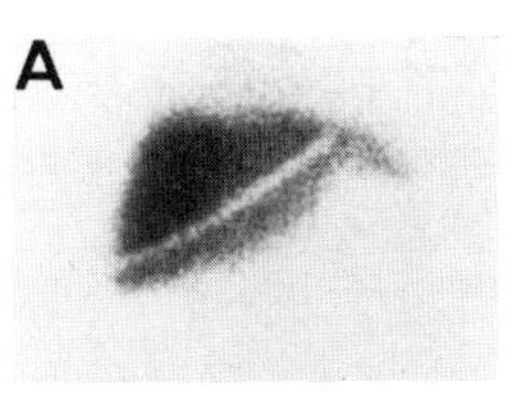

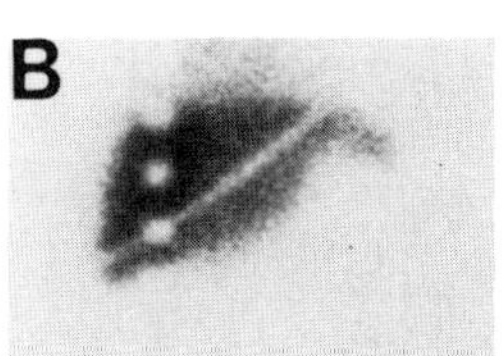

Fig. 3. A. Identification of right costal margin. B. Longitudinal measurement of liver image performed in the anterior position. The measuring device consists of 2-cm lead markers alternating with 2-cm blank areas.

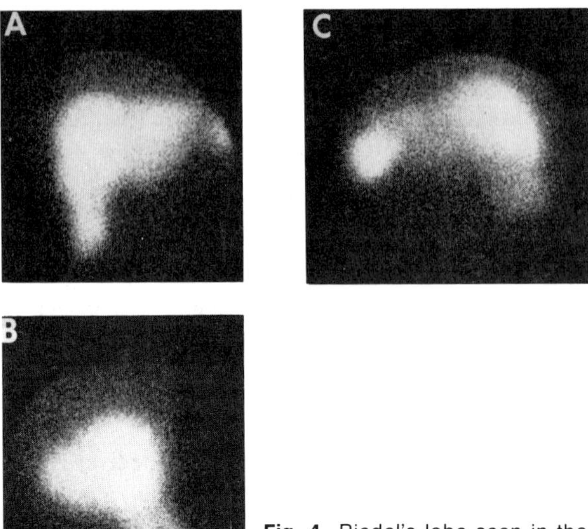

Fig. 4. Riedel's lobe seen in the anterior (A), right lateral (B), and posterior (C) positions.

patients the organ may be made up of only a single lobe, while in others four or five lobes may be identified. Unusual configuration in a normal liver may result in mistaken interpretation of the liver scan.

It is not uncommon for a Riedel's lobe to be present, a finding seen in approximately 10 percent of cases. This supernumerary lobe usually appears as an extension of the inferior portion of the right lobe, is often globular in shape, and is well seen in both the anterior and right lateral positions (Fig. 4). Riedel's lobe may be anterior or posterior in location.

The palpable liver is often not enlarged. This is illustrated in Figure 5, in which a normal-sized, low-lying liver extends well below the right costal margin. Elevation of the right hemidiaphragm may be reflected on the liver scan as hepatic tissue extends superiorly (Fig. 6).

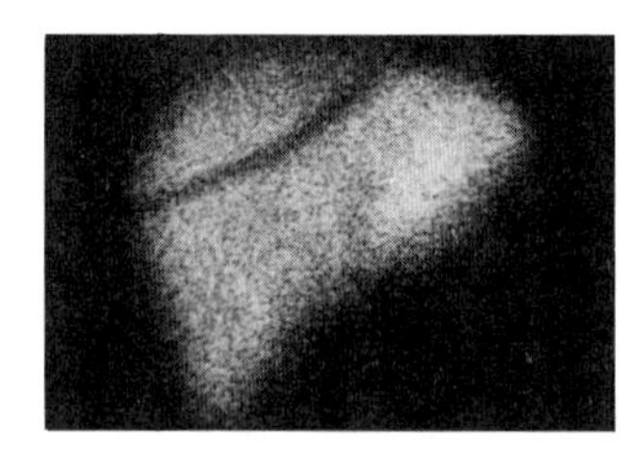

Fig. 5. Normal-sized, low-lying liver.

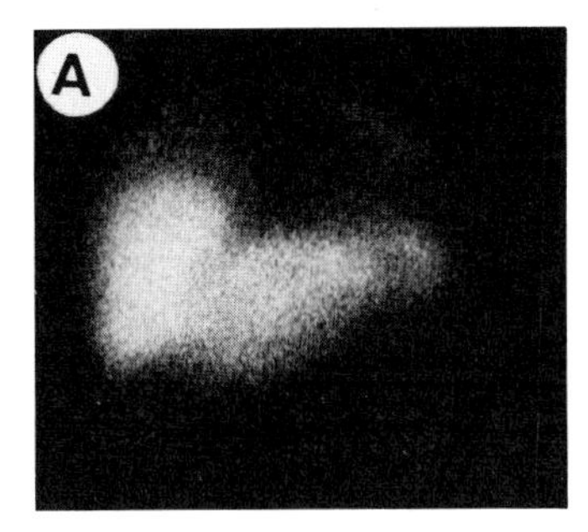

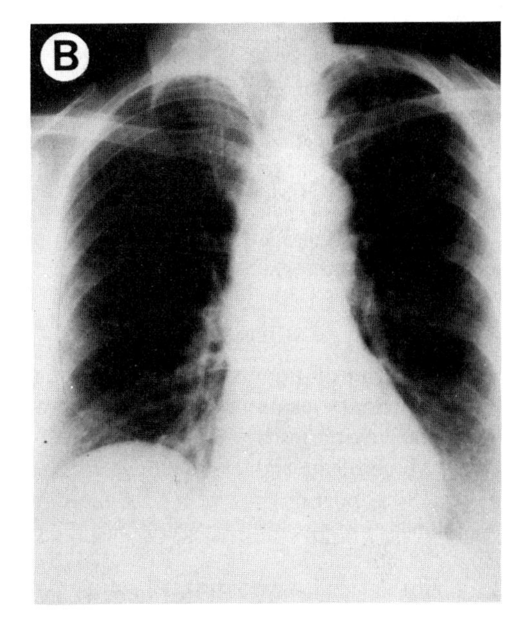

Fig. 6. Altered configuration of the subdiaphragmatic portion of the right lobe in a case of relative elevation of the right diaphragm. A. The upward extension of the right lobe is a reflection of the diaphragmatic position seen on the chest x-ray in B.

Abnormal Liver Imaging

If hepatic tissue with its accompanying Kupffer cells has been replaced or destroyed, the radioactive colloid is unable to enter the involved region and thus is represented on the scan as an area of absent radioactivity. This is the basis for liver scan visualization of space-occupying lesions.

Most liver scan abnormalities involve identifying such areas of absent radioactivity. In addition, the abnormal liver scan may detect hepatic displacement or distortion of hepatic architecture. It is convenient to classify liver scan abnormalities as localized absent radioactivity or diffuse reduction in radioactivity.

Localized Areas of Absent Radioactivity. A space-occupying lesion in the liver may be visualized on the hepatic scan as a localized area of absent radioactivity. Although neoplasm is most commonly responsible for this type of defect, the scan appearance of other kinds of space-occupying lesions *may be identical.* Thus it is often impossible to distinguish between tumor, cyst, abscess, or hematoma. The most commonly encountered space-occupying lesion seen on a liver scan is metastic tumor.

Metastatic Tumor. The image of a solitary liver metastasis (Fig. 7) shows the lesion to be well delineated and easily identified. In this case the liver is neither enlarged nor distorted.

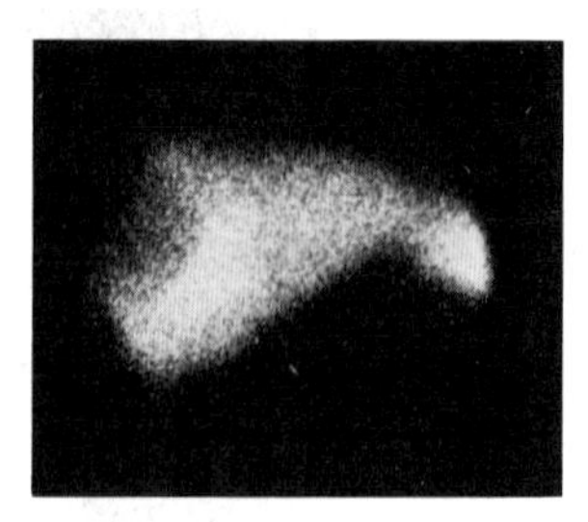

Fig. 7. Localized area of absent radioactivity, indicating a solitary metastasis.

The scan of a patient with extensive metastatic involvement of the liver from a primary lesion in the breast is shown in Figure 8. Note the numerous space-occupying lesions and the marked enlargement and distortion of the liver. Much of the right lobe has been replaced by tumor, and the remaining, functioning portion has been displaced medially. Other examples of liver metastases are given in Figure 9, and the progression of metastases in Figure 10.

Although the majority of hepatic metastases may be demonstrated on the anterior scan alone, scanning in the right lateral position is often helpful. This is especially true in lesions involving the subdiaphragmatic portion of the right lobe or the posterior aspect of the liver. In Figure 11 the metastatic deposit is clearly delineated on the right lateral scan, whereas this is not so on the scan done in the anterior position. In other situations the major findings may be visualized best in the posterior position.

Metastatic lesions are most often noted in the right lobe, due in part to

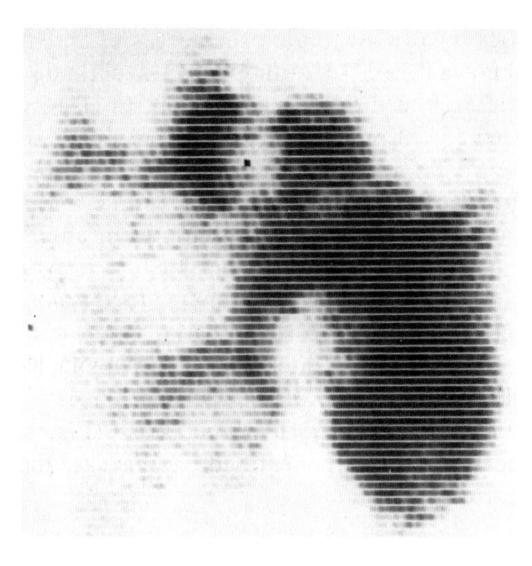

Fig. 8. Enlarged, distorted liver with multiple, round space-occupying lesions due to metastases from carcinoma of the breast.

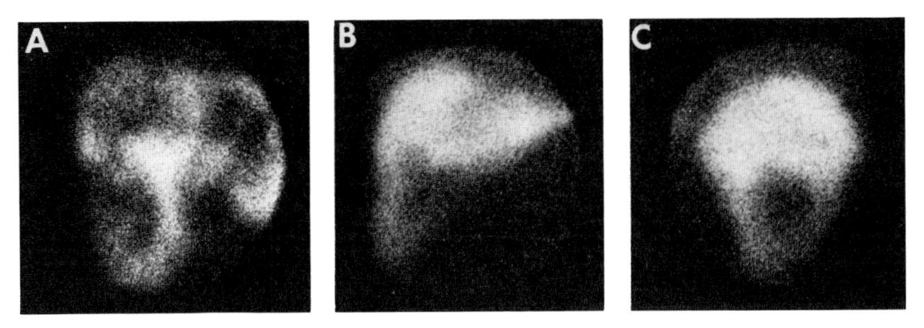

Fig. 9. Liver metastases. A. Anterior hepatic image showing extensive neoplastic involvement. B and C. Enlarged liver showing metastatic involvement in both anterior (B) and right lateral (C) positions.

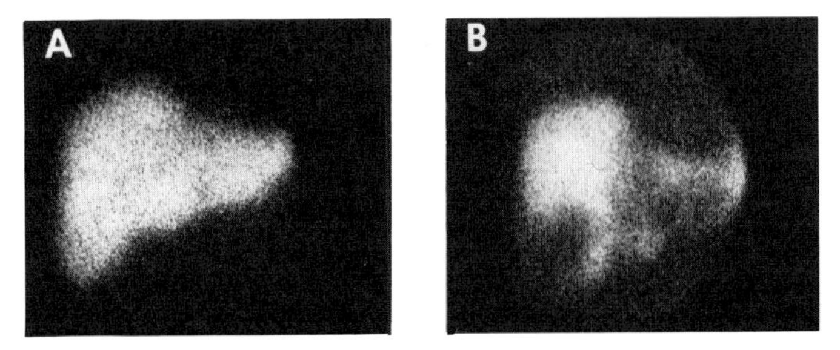

Fig. 10. Progression of metastases imaged over a period of 1 year. A. No discernible space-occupying lesions. B. Extension of metastases and alteration of hepatic configuration.

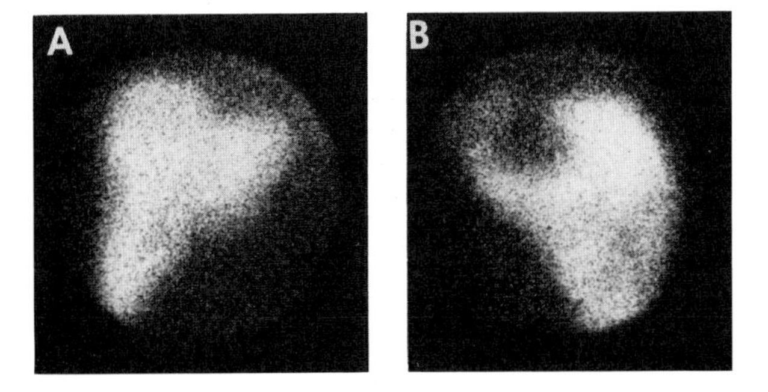

Fig. 11. Usefulness of the right lateral scan. Although there are several metastases seen in the anterior position (A) the lesion in the subdiaphragmatic portion of the right lobe can be well appreciated only in the right lateral position (B).

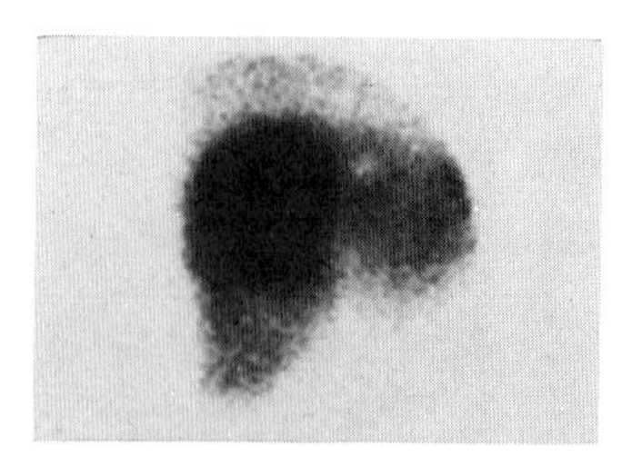

Fig. 12. Small solitary metastatic lesion seen in the left lobe imaged in the anterior position.

its relatively larger size. However, metastases are sometimes demonstrated in the left lobe only (Fig. 12).

Primary Liver Tumors. A primary liver tumor has much the same scan appearance as a single metastatic lesion. As might be expected, resection of portions of hepatic tissue or removal of an entire lobe results in unusual imaging of the remaining liver. For example, in the case shown in Figure 13 the right lobe, which was involved with tumor, was removed. The remaining, visualized portion has undergone compensatory hypertrophy. Without the history of surgical intervention such a scan might be mistakenly thought to represent unusual medial displacement of the liver.

Nonneoplastic Space-Occupying Lesions. As a rule, most space-occupying lesions of the liver have similar scan appearances, and it is usually difficult or impossible to distinguish between those due to tumor and those

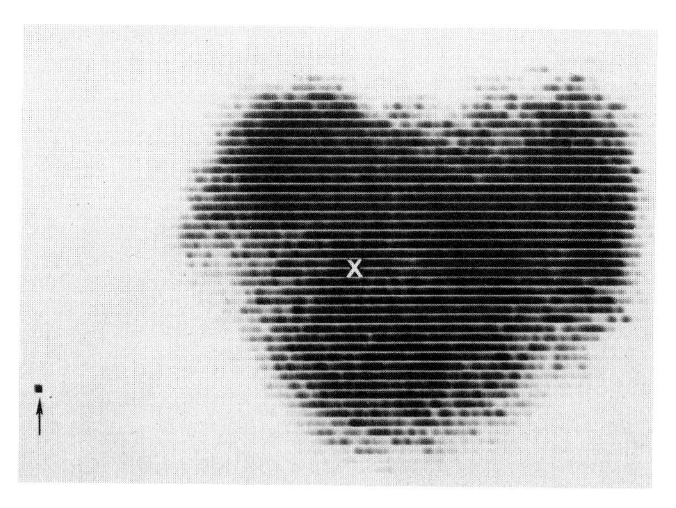

Fig. 13. Anterior scan showing hepatic regeneration in patient in whom the right lobe was removed because of neoplasm. Note the position of the liver in relation to the xiphoid (X) and the right lateral body wall (arrow).

that are nonneoplastic. The latter includes such entities as cyst, abscess, hematoma, nodular cirrhosis, and extrinsic pressure defect from an adjacent organ or mass. In each of these situations, as in tumor, the underlying disorder has resulted in the replacement or destruction of hepatic tissue and accompanying Kupffer cells. In those instances in which the lesion is well circumscribed and localized, the scan appearance of the abnormality is identical to that seen with neoplasm, both primary and metastatic.

Cystic lesions of the liver do not have a characteristic appearance and may be variable in size and location. The identification of echinococcus (hydatid) cysts of the liver with radionuclide imaging procedures is not unusual.

It is now well appreciated that marked dilatation of the biliary system, as a consequence of extrahepatic biliary obstruction, may be visualized with radionuclides. The abnormality is usually seen as a stellate-shaped area of absent radioactivity (Fig. 14).

For purposes of radionuclide imaging, abscesses of the liver may be categorized into three types: intrahepatic, subdiaphragmatic, and subhepatic. Intrahepatic abscesses may be single or multiple, and the possibility of confusing such lesions with neoplasm on liver scans is well known. A solitary abscess of the right lobe is seen in Figure 15.

A right-sided subdiaphragmatic abscess is sometimes difficult to demonstrate on the liver scan alone. Such a lesion is often best recognized on a combined lung-liver scan.[1] Because the inferior portion of the right lung (as it overlies the diaphragm) envelops the subdiaphragmatic portion of the right lobe of the liver, a purulent collection beneath the diaphragm causes some encroachment on both the liver and lung images displayed on the scan.

When performing a combined lung-liver scan, it is convenient to choose (1) the same tracer for imaging each of the organs or (2) one radionuclide for visualizing the lung and another for the liver, with the two radioactive materials having similar energies. Thus if the same tracer is to be used, ^{99m}Tc-

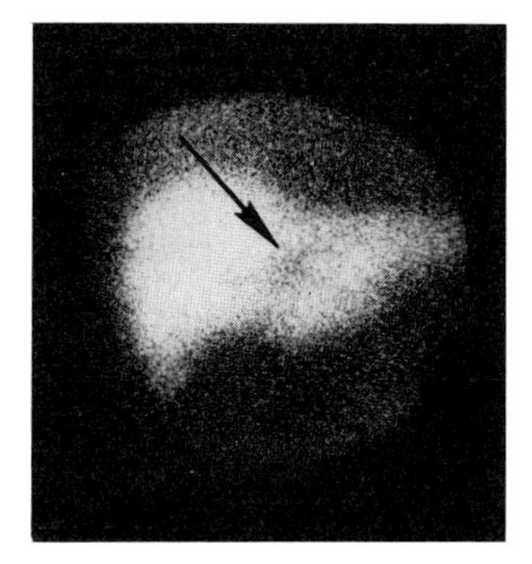

Fig. 14. Stellate shaped area of absent radioactivity associated with marked biliary dilatation in a patient with obstructive jaundice.

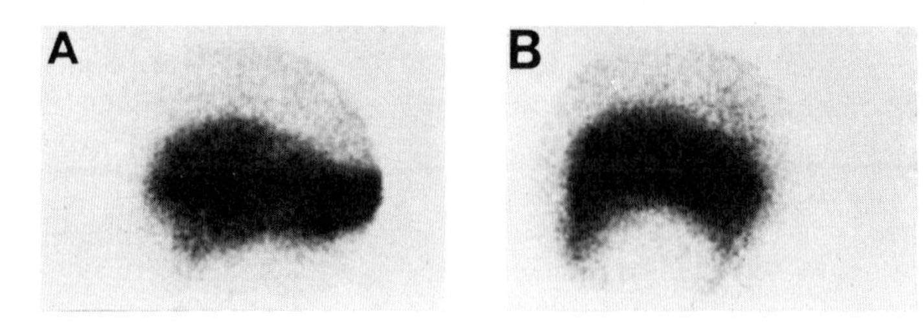

Fig. 15. Anterior (A) and right lateral (B) images in a patient with amebic abscess of the right lobe. Note the thin wall of the lesion as seen in the right lateral position.

albumin microspheres may be selected for the lung and ^{99m}Tc-sulfur colloid for the liver. On the other hand, if two different radionuclides are used ^{131}I-MAA may be employed for the lung and ^{198}Au-colloidal gold for the liver.

A normal lung-liver scan is shown in Figure 16. Note that there is no evidence of reduced radioactivity between the right lung and the liver. Indeed, maximum radioactivity is noted in the region in which the lung envelops the right hepatic lobe.

A combined lung-liver scan demonstrating a right-sided subdiaphragmatic abscess is shown in Figure 17A. In this case the collection extends laterally and inferiorly. Although the liver scan alone offers a hint to the diagnosis (Fig. 17B), the abnormality is much more sharply outlined on the combined study.

If a pleural effusion or other pulmonary pathology adjoining the liver is present, the usefulness of this procedure is greatly restricted. Unfortunately,

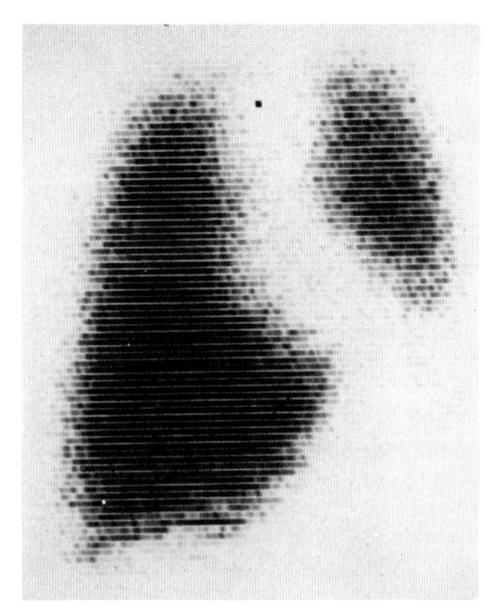

Fig. 16. Normal anterior lung-liver scan.

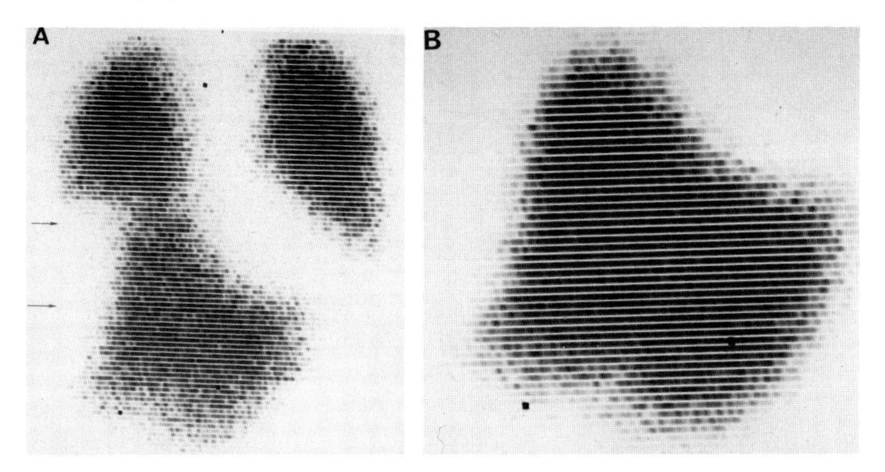

Fig. 17. A. Anterior lung-liver scan demonstrating subdiaphragmatic abscess (arrows). B. Anterior liver scan of same patient in which abscess would be more difficult to identify.

left-sided subdiaphragmatic collections are not so well visualized as those related to the liver.

Not uncommonly, adjacent organs or masses may cause extrinsic pressure on the liver resulting in distortions on the liver scan that may be indistinguishable from hepatic space-occupying lesions. This is often seen with an enlarged kidney or neoplasm in a structure closely related to the liver.

Hematomas involving the liver may appear as extrinsic or intrinsic lesions. A subdiaphragmatic hematoma secondary to trauma, causing extrinsic compression of the liver is seen in Figure 18. Note how round the defect is. Following drainage, the liver scan returned to normal. Intrahepatic hematomas are sometimes seen in patients who have been treated with anticoagulants.

Diffuse Liver Disease. One of the most informative presentations in radionuclide imaging is that of the liver affected by a diffuse disease process, such as Laennec's cirrhosis. This type of disorder usually involves the entire organ and is so represented on the scan.

It can be generally stated that where there has been parenchymal cell destruction or replacement by fibrosis associated with cirrhosis, there is also similar replacement of the Kupffer cells. On the other hand, parenchymal cell regeneration is also accompanied by a regeneration of Kupffer cells. Thus both types of cells may serve to reflect the status of the liver when cirrhosis is present.

During the early stages of cirrhosis the histologic picture is characterized by areas of fibrosis (with hepatic tissue replacement) and other areas of liver regeneration. The scan of such a liver exhibits uptake in the unaffected and

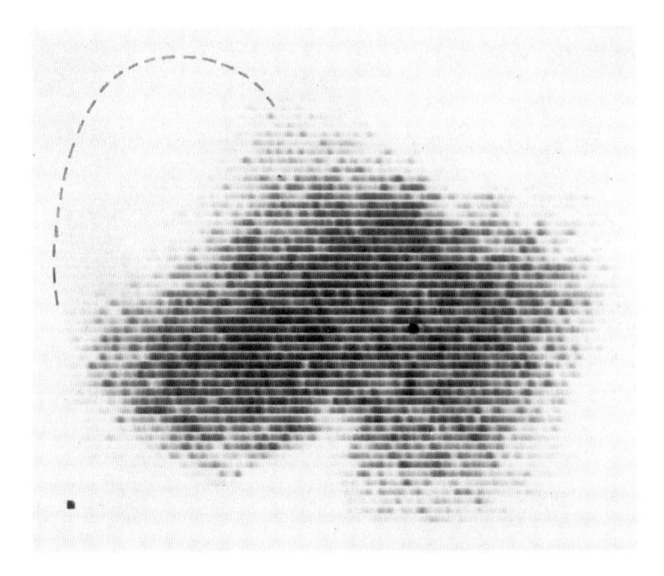

Fig. 18. Subdiaphragmatic hematoma. The normal hepatic configuration is indicated by the broken line.

regenerated portions of the organ and areas of absent activity where there has been fibrous tissue growth (Fig. 19). The picture of alternate areas of increased and reduced (or absent) radioactivity is simply a representation of the liver's pathologic state. This is sometimes referred to as radioactive mottling.

Since the liver is not able to accept the same amount of radiocolloid as it would if there were no hepatic disease, some of the rejected radionuclide becomes phagocytized in the spleen and this organ becomes more prominent in both degree of radioactivity and size. With progression of the disorder, there is combined fatty metamorphosis and increased fibrous tissue replacement. This results in the liver being less able to phagocytize the radioactive colloid, and the rejected radioactive material is taken up elsewhere in the RES. At this stage the amount of radiocolloid presented to the spleen even exceeds what can be phagocytized in that organ, and the excess radionuclide becomes engulfed in the RES cells of the bone marrow. Although all the marrow is involved, it is common to visualize the bone marrow in the thoracic region because of the superimposition of radioactivity in the sternal and vertebral marrow. As the cirrhosis becomes even more advanced there is continued loss of the liver's ability to phagocytize the radioactive colloid so that very little hepatic uptake is seen on the scan and the spleen and bone marrow become unusually prominent. When ascites is present, the liver may appear to be displaced medially.

Distinguishing extensive neoplastic involvment of the liver from far ad-

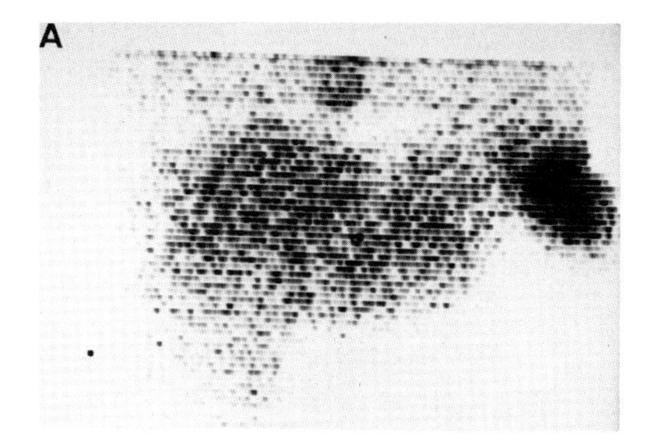

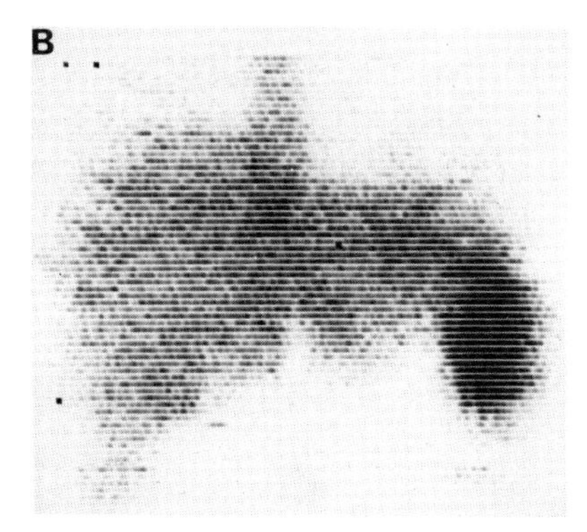

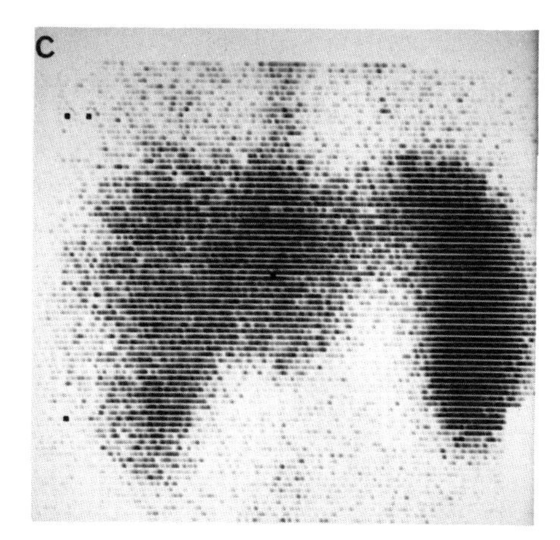

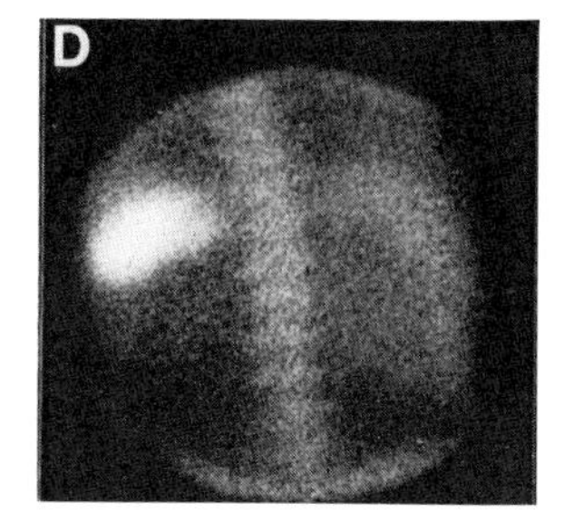

Fig. 19. Liver scans in patients with Laennec's cirrhosis. A and B. Radioactive mottling of liver, increased splenic radioactivity, and radioactive material in the bone marrow. C. The spleen shows relatively increased radioactivity and is markedly enlarged. D. Posterior image shows only minor hepatic uptake, prominence of the spleen, and radioactive material in the bone marrow.

vanced cirrhosis may sometimes pose a problem. If hepatic metastases are present the portion of liver not involved with tumor often shows normal liver function and the uninvolved region may accept all the radiocolloid presented to it with subsequent failure to image distant RES structures, such as the bone marrow. On the other hand, since entities such as cirrhosis involve the whole liver without sparing certain portions, rejected radioactive colloid results in the imaging of other portions of the RES.

Large localized areas of hepatic tissue destruction are often seen on the liver scan of the cirrhotic patient. Sometimes this may represent a hepatoma, especially if the lesion is rounded. The incidence of hepatoma rises in patients with cirrhosis (Fig. 20). In still other patients rounded space-occupying lesions may merely be nodular cirrhosis, and the scan may look exactly like that of neoplasm.[2]

From the standpoint of radiocolloid imaging, other diffuse hepatic diseases behave in a fashion similar to Laennec's cirrhosis. These may include such maladies as tuberculosis, sarcoidosis, and amyloidosis.

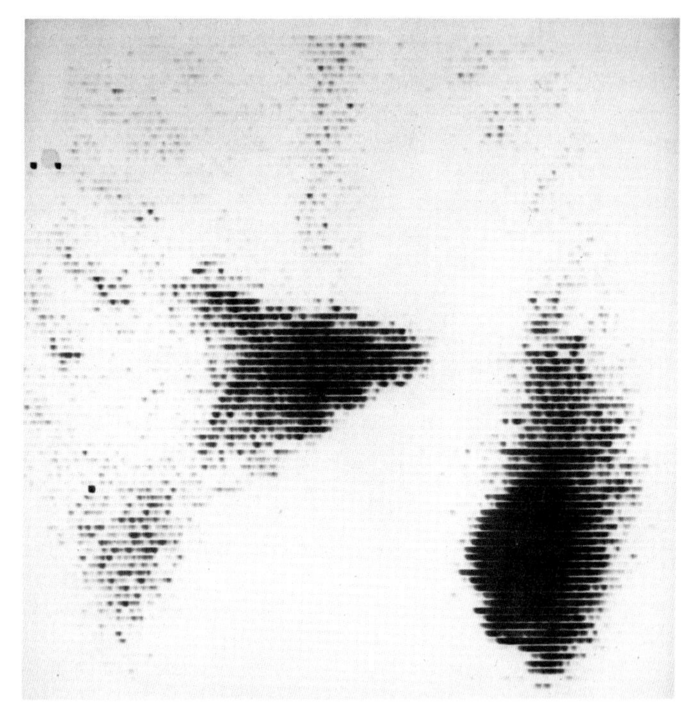

Fig. 20. Anterior scan of patient with Laennec's cirrhosis who developed a hepatoma, the latter seen as the rounded area of absent radioactivity in the right lobe. Note the radioactivity in the costal marrow.

Radioiodinated Rose Bengal Studies

Following intravenous administrations of ^{131}I-rose bengal, the radionuclide is taken up by the parenchymal cells of the liver and excreted into the biliary system. The presence of the radionuclide in the liver provides a means to scan the organ, and the excretion (or lack thereof) may be utilized to study hepatic function.[3]

Scanning. Liver scanning may commence 15 to 30 minutes following radionuclide injection of 150 to 300 μCi (Fig. 21). However, after this time radioactive material is seen in the small intestine as the ^{131}I-rose bengal is excreted via the biliary system into the duodenum (Fig. 22). Failure to recognize radionuclide in the small intestine may result in wild misinterpretations of the liver scan. By 2 hours after administration of the radioactive material, its presence is easily discerned in segments of small bowel.

An inconstant finding is the scan imaging of the gallbladder as the radioiodinated rose bengal is excreted into this organ (Fig. 22C). Gallbladder visualization may be seen 2 to 6 hours after radionuclide injection in the normal subject.

Because ^{131}I-rose bengal is normally taken up only by the hepatic parenchymal cells and does not enter the phagocytic cells of the RES, there is no splenic or bone marrow visualization on the scan. This property may be utilized in situations in which portrayal of the hepatic image only is desired. Thus if situs inversus is suspected, the presence of a left-sided organ can be confirmed to be the liver because of the appearance of rose bengal in the structure.

If there is a reduction in the excretory ability of the liver—whether from hepatocellular disease or extrahepatic biliary obstruction—these changes may be represented on the rose bengal scan by a failure to visualize excreted radionuclide in the small bowel. Because an absence of radioiodinated rose bengal in the small intestine may be found in both extra and intrahepatic

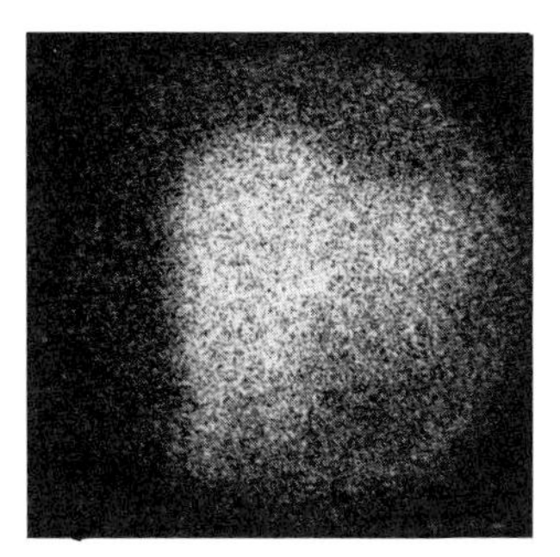

Fig. 21. Normal hepatic imaging performed 15 minutes after injection of ^{131}I-rose bengal.

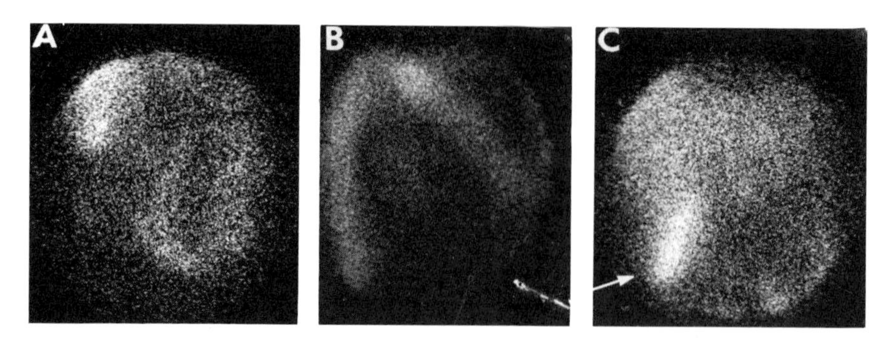

Fig. 22. A. Evidence of radioactivity in the small intestine 2 hours after radioiodinated rose bengal administration. B. At 24 hours after injection. C. Visualization of the gallbladder (arrow) 2 hours after tracer administration.

biliary obstruction, the scan can not itself distinguish these two entities.

Several problems may occur when attempting to image the liver of a jaundiced patient. Since the excretory ability of the liver is greatly impaired, scan visualization may be unrewarding unless a large quantity of radioactive material is administered. Rejected radionuclide that cannot be removed from the blood by the diseased parenchymal cells may become excreted via the kidneys, and a renal image may be seen on the scan, sometimes leading to confusion in organ identification.

Function Studies. External monitoring studies can be used to estimate hepatic excretion of ^{131}I-rose bengal.[4] For this procedure a probe is placed over the ear to monitor the radioactivity in the circulatory system, and another is placed over the left paraumbilical area to detect radioactivity in the intestinal tract.

Radioactivity at each site is determined at 5 and 30 minutes after radionuclide injection. Because the 5-minute value over the ear represents radioactivity after uptake of the ^{131}I-rose bengal by the parenchymal cells:

$$\frac{\text{blood radioactivity at 30 minutes}}{\text{blood radioactivity at 5 minutes}} \times 100 = \text{percent radionuclide remaining in the blood at 30 minutes after injection}$$

Normally this is no greater than 40 to 50 percent. If more than 50 percent is present after 30 minutes, there is a delay or defect in the ability of the polygonal cells to extract radionuclide from the blood. In both intra- and extrahepatic biliary obstruction very little is cleared from the blood in this short time.

The abdominal probe reflects in part the discharge of ^{131}I-rose bengal into the small intestine. In cases of biliary obstruction of both varieties there is scant or no increase in radioactivity over the period of 5 to 30 minutes. In some situations it is helpful to administer a fatty food such as milk to help stimulate biliary excretion through hormonal release (cholecystokinin). Such stimulation usually does not result in increased radioactivity in the intestine in cases of extrahepatic biliary obstruction, whereas in some instances of hepatocellular disease some radioactive material may reach the intestinal tract.

^{131}I-rose bengal studies have been useful in detecting congenital biliary atresia in infants. A dose as small as 5 μCi may be used in this age group.

Further Clinical Considerations

In many instances an accurate diagnosis may be forthcoming from the liver scan alone. This is frequently encountered in patients with Laennec's cirrhosis or extensive neoplastic involvement of the liver. However, scan interpretation must often be tempered by the knowledge that most space-occupying lesions of the liver have a similar scan appearance. It is surprising how unrelated entities such as metastatic tumor and nodular cirrhosis look alike on a liver scan.[2]

Scan Accuracy. One of the prime concerns in evaluating the patient suspected or known to have a malignancy is whether there is metastatic involvement of the liver. To be sure, hepatic scanning is probably the single most valuable screening procedure to help make this determination, even though it has limitations

The accuracy of liver scans interpreted as showing evidence of space-occupying lesions (e.g., tumor) is approximately 80 to 90 percent[5]—i.e., 10 to 20 percent of those patients whose scans were thought to have evidence of lesions such as metastases are ultimately found to be free of such disease! This is an alarming statistic but must be reckoned with in those situations in which knowledge of the presence or absence of liver metastases is essential, especially in deciding whether radical or only palliative surgery should be performed.

It is sometimes difficult to conceive that neoplasm is absent when the liver scan may suggest the presence of one or more space-occupying lesions. The scan shown in Figure 23 was interpreted as demonstrating a space-occupying lesion in the right lobe with upward displacement of the liver. In reality the liver was found to be free of tumor and consisted of a rudimentary right lobe closely related to the superior and posterior subdiaphragmatic surfaces, and a prominent left lobe. Of course in this situation a lateral scan can help clarify the suspected abnormality. On the other hand, the problem of inaccuracy is greatly reduced in those cases in which there is multiple metastatic involvement of the liver.

Liver Biopsy. If one or more space-occupying lesions has been identified through radionuclide imaging, the scan may be extremely useful in locating a precise site at which a closed needle biopsy may be made. Should the scan

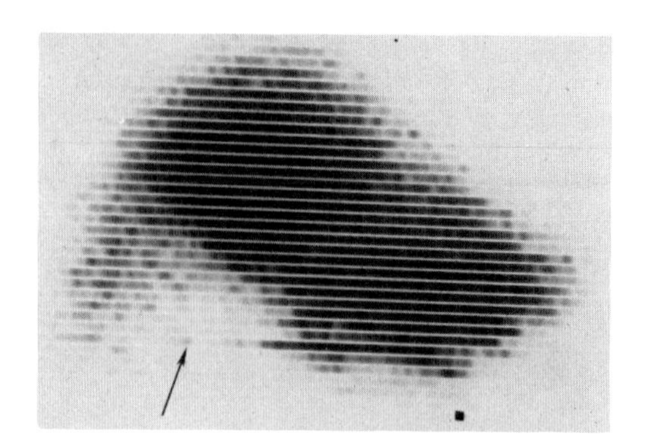

Fig. 23. Anterior liver scan in which a space-occupying lesion (arrow) was thought to be present. At surgery the liver was found to be free of tumor. The dots outline the right costal margin.

demonstrate a lesion inaccessible to closed biopsy, then open biopsy should be considered.

Response to Chemotherapy. The behavior of liver metastases following cancer chemotherapy offers objective evidence of a favorable or unfavorable response to the specific mode of treatment. These changes include the size of both the liver and the involved tumor.

SPLEEN

Radionuclide studies of the spleen have been involved with both organ imaging and evaluation of splenic function. The latter is covered in more detail in the section on hematologic procedures in Chapter 7.

Currently spleen scanning may be accomplished by two general methods: (1) phagocytosis of a radiocolloid by the spleen's RES cells or (2) sequestration in the spleen of heat-damaged erythrocytes tagged with a tracer. The utilization of radioactive colloids is the method most commonly employed.

Imaging with Radioactive Colloids

If a radioactive colloid (such as that with a particle size of approximately 0.1μ in diameter) is injected intravenously, the material is phagocytized first by the endothelial lining Kupffer cells of the liver. When the Kupffer cells become "saturated" the excess colloid is then available for uptake in the spleen through phagocytosis by the macrophages of the splenic pulp. The remaining radioactive colloid may enter the phagocytic cells of the RES in the bone marrow. It is convenient to remember the hierarchy of liver to spleen to bone marrow in regard to phagocytic activity.

When using a radioactive colloid for spleen scanning it is essential that the liver is first overloaded with radionuclide so that a sufficient amount remains for splenic uptake. This is almost always accomplished by 1 mCi ^{99m}Tc-sulfur colloid. For splenic imaging with ^{198}Au-colloidal gold, doses in the range of 300 to 400 μCi are necessary. Unless otherwise stated, all reference here to radiocolloid imaging of the spleen involves the use of ^{99m}Tc-sulfur colloid as the scanning agent.

The optimum position for scanning the spleen may vary of course with the organ's location. Although in most cases the organ has a relatively posterior location, in some situations it may lie at various planes in the body, from posterior to anterior. An excellent method to determine its relative anteroposterior position is to perform a right lateral scan of the liver first; this assists in further delineating the spleen's location. Sometimes it is preferable to image the spleen in a modified left oblique position.

Normal images of the spleen are shown in Figure 1. The precise size and position of the organ can be readily determined with the scanning procedure. As with liver scanning, lead markers may be placed on the skin over the splenic area, anterior or posterior, so the organ's size and position as related to the costal margin may be determined. It is often useful to image the spleen in both anterior and posterior positions despite the organ's location. Sometimes its longitudinal axis can not be appreciated or well determined unless this is done. Variants of splenic position may be seen with radionuclide imaging, such as relative elevation of the organ (Fig. 24).

Splenomegaly is the most frequently encountered abnormality in spleen scanning (Fig. 25A and B). The spleen normally measures no more than 12 cm at its greatest longitudinal dimension. Countless disorders may be associated with an enlarged spleen, although occasionally splenomegaly may be an isolated finding with no demonstrable underlying disease. Most often, a scan showing an enlarged spleen indicates cirrhosis. The spleen is often enlarged also in patients with lymphoma or leukemia, and its subsequent reduction or enlargement on the scan may offer objective evidence of the patient's response to a specific chemotherapeutic agent.

Another interesting finding observed in lymphoma is a relative increase

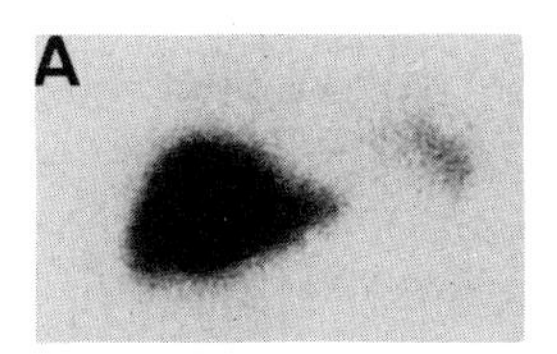

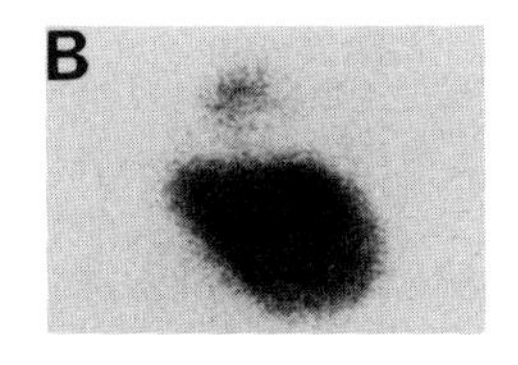

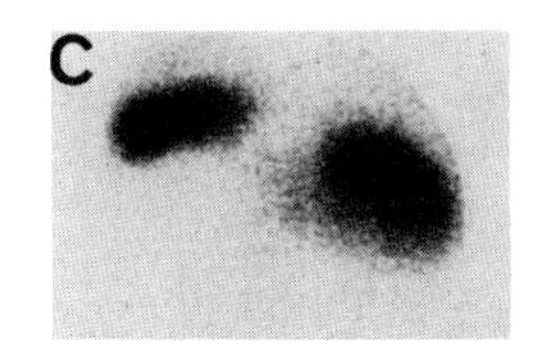

Fig. 24. Relative elevation of the spleen seen in the anterior (A), right lateral (B), and posterior (C) positions.

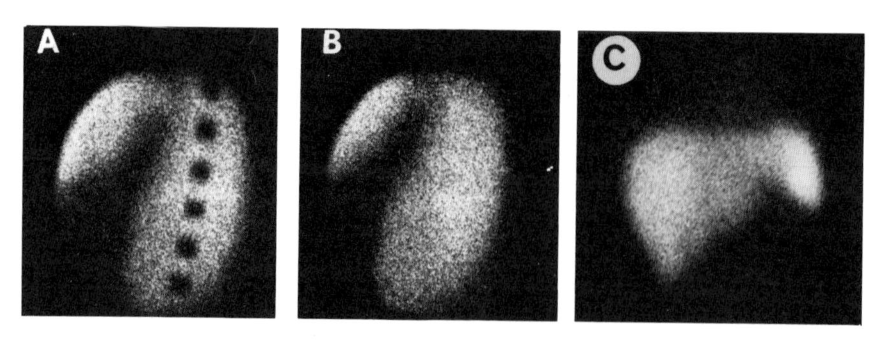

Fig. 25. Common splenic abnormalities. Enlarged spleen in a patient with infectious mononucleosis. Scan performed in the anterior position with (A) without (B) 2 cm markers. C. Imaging in the anterior position shows a relative increase in splenic radioactivity, as compared to the liver, in a patient with Hodgkin's disease.

in radioactivity in the spleen, when compared to the liver, in the absence of known or suspected hepatic involvement (Fig. 25C). The increased splenic activity may be a manifestation of heightened phagocytic ability of splenic macrophages in lymphomas.

Spleen scans of a patient with acute myelogenous leukemia before and after treatment are shown in Figure 26. Note the marked reduction in size of the organ following chemotherapy.

Identification of left upper quadrant masses is greatly facilitated with splenic scanning. Whether a mass in this area is related to the spleen can be determined when comparing the scan and the physical and roentgenographic findings.

Space-occupying lesions have also been demonstrated on spleen scan-

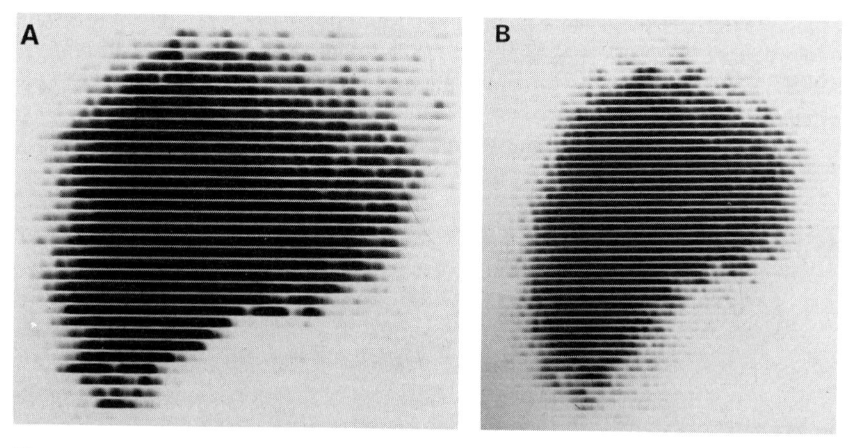

Fig. 26. Posterior spleen scans in a patient with acute myelogenous leukemia before (A) and after (B) chemotherapy.

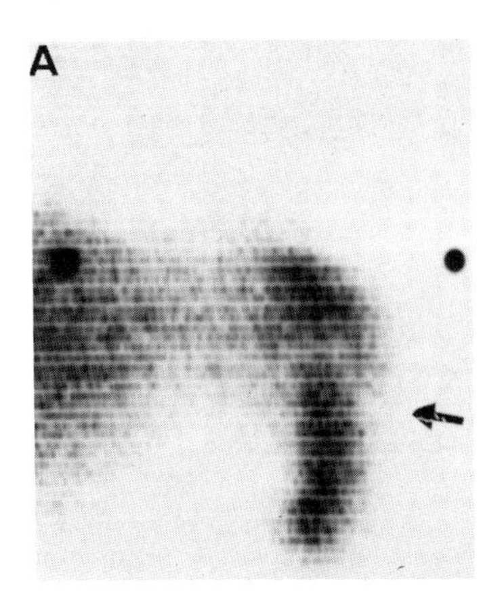

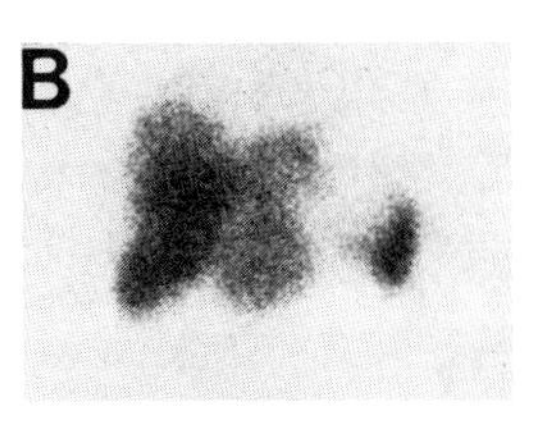

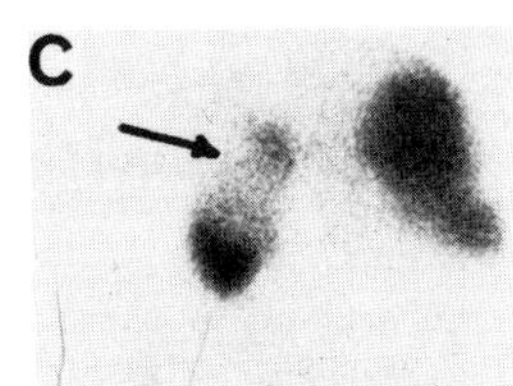

Fig. 27. Trauma to the spleen. A. Anterior scan showing area of absent radioactivity (arrow) due to laceration of the spleen. (From Dr. George W. Evans et al: JAMA 1973.) B and C. Absent radioactivity associated with an old, encapsulated splenic hematoma (arrow) seen in the (B) anterior and (C) posterior positions.

ning. Among these entities are abscess, lymphomatous involvement, and granulomatous lesions of sarcoid. Such involvement is nonspecific, and the etiology cannot be established from the scan.

Evidence of splenic infarction can be seen on the scan. This usually appears as a segmental area of absent radioactivity.

Trauma to the spleen may sometimes be demonstrated through radionuclide imaging. Evidence of splenic lacerations and subcapsular hemotomas have been seen (Fig. 27).

The term "functional asplenia" has been used to describe the failure to image the spleen during a crisis of sickle cell anemia.[6] The inability to visualize the spleen at such times is thought to be due to the failure to deliver the radionuclide to the spleen's macrophages because there is reduced perfusion to the spleen; the latter is due to elevated blood viscosity during the sickle cell crisis. Following resolution of the crisis and resultant improvement in splenic artery perfusion, a sufficient quantity of radionuclide can then be delivered to the spleen (Fig. 28).

Imaging with Tagged, Damaged Erythrocytes

It is well known that certain disorders (e.g., some hemolytic anemias) are characterized by coated or damaged erythrocytes that become sequestered in the spleen. If a small volume of a patient's cells are withdrawn, damaged by heating, and reinjected, they are sequestered by the spleen too.[7] If, in addition, these damaged erythrocytes are tagged with a gamma emitter, the spleen may be visualized through radionuclide imaging.

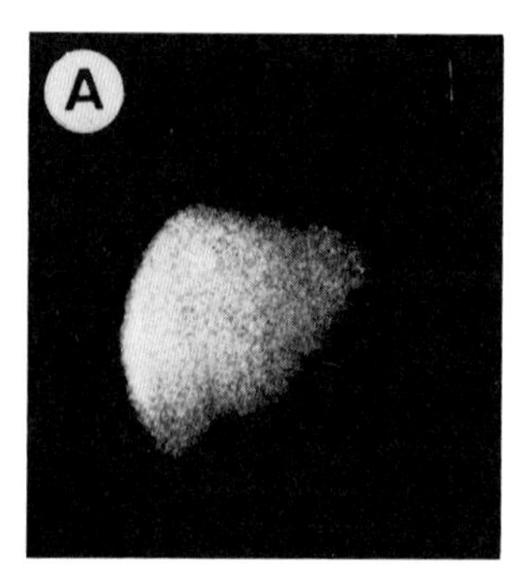

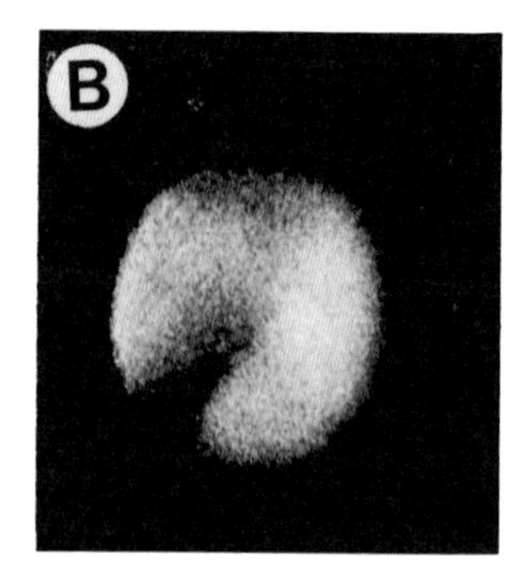

Fig. 28. Functional asplenia in a patient with sickle cell-hemoglobin C disease and splenomegaly. A. Practically nonvisualization of the spleen during crisis. B. Imaging of the enlarged spleen following resolution of crisis, when radionuclide could be delivered to organ. (From Joshpe et al: Am J Med 55:720, 1973)

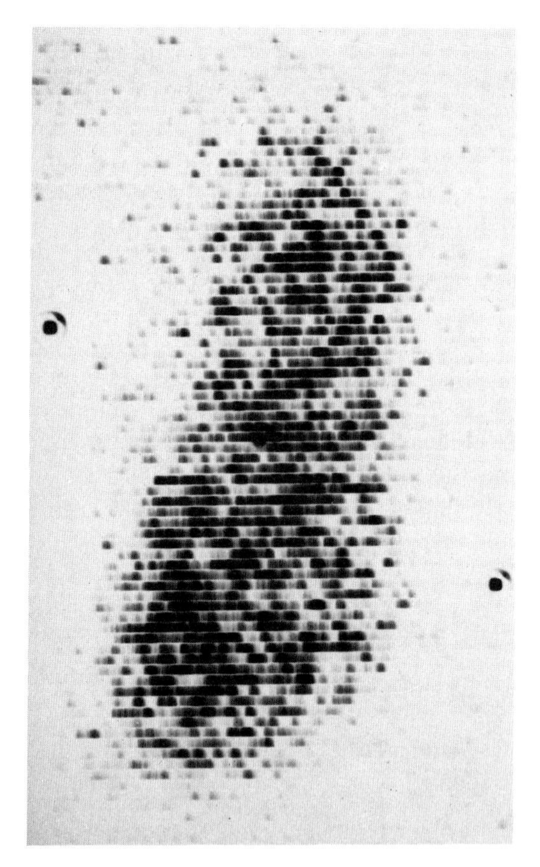

Fig. 29. Spleen scan performed in the anterior position with heat-damaged erythrocytes tagged with ^{51}Cr-sodium chromate.

The procedure for labeling erythrocytes with ^{51}Cr-sodium chromate is described in Chapter 7. For spleen scanning, approximately 300 μCi ^{51}Cr are used. The tagged cells are damaged by heating them for 1 hour at 50 C. The tracer solution is allowed to cool to room temperature, the material is injected intravenously, and the patient may be scanned 2 to 24 hours after radionuclide administration.

By using the tagged erythrocyte method, only the spleen is visualized and there is no interference from other structures in the RES, as is encountered in scanning with radioactive colloids. The spleen is well seen on the ^{51}Cr scan, and its dimensions are readily identified (Fig. 29).

Splenic scanning by this method is useful when there is confusion in identifying accessory spleens because such tissue is sometimes difficult to distinguish from liver. Of course, each of the splenic entities described in the section on colloid scanning of the spleen may also be demonstrated by the ^{51}Cr method.

REFERENCES

1. Brown DW: Lung-liver scans in the diagnosis of subdiaphragmatic abscess. JAMA 197:728, 1966
2. Klion FM, Rudavsky AZ: False-positive liver scans in patients with alcoholic liver disease. Ann Intern Med 69:283, 1968
3. Whiting EG, Nusynowitz ML: Radioactive rose bengal testing in the differential diagnosis of jaundice. Surg Gynecol Obstet 127:729, 1968
4. Nordyke RA, Blahd WH: Blood disappearance of radioactive rose bengal. JAMA 170:1159, 1959
5. Ludbrook JL, Slavotinek AH, Ronai PM: Observer error in reporting on liver scans for space-occupying lesions. Gastroenterology 62:1013, 1972
6. Pearson HA, Spencer RP, Cornelius EA: Functional asplenism in sickle-cell anemia. N Engl J Med 281:923, 1969
7. Winkelman JW, Wagner HN Jr, McAfee JG, et al: Visualization of the spleen in man by radioisotope scanning. Radiology 75:465, 1960

ADDITIONAL READING

DeLand FH, Wagner HN Jr: Reticuloendothelial system, liver, spleen and thyroid. In: Atlas of Nuclear Medicine, vol III. Philadelphia, Saunders Co., 1972, pp 63-233.

Freeman LM, Blaufox MD (eds): Radionuclide Studies of the Gastrointestinal System. New York, Grune and Stratton, 1973

Johnson PM: The liver. In Freeman LM, Johnson PM (eds): Clinical Scintillation Scanning. New York, Hoeber Medical Division, Harper and Row, 1969, pp 260-303

McAfee JG, Ause RG, Wagner HN Jr: Diagnostic value of scintillation scanning of the liver. Arch Intern Med 116: 95-110, 1965

Rosenthall L: The Application of Radioiodinated Rose Bengal and Colloidal Radiogold in the Detection of Hepatobiliary Disease. St. Louis, Warren H. Green, Inc., 1969

Schwabe AD: Gastrointestinal tract function and disease. In Blahd WH (ed): Nuclear Medicine, 2nd ed. New York, McGraw-Hill, 1971, pp 350-366

Shingleton WH: Liver scanning. In Blahd WH (ed): Nuclear Medicine, 2nd ed. New York, McGraw-Hill, 1971, pp 366-394

Wagner HN Jr, Mishkin F: The liver. In Wagner HN Jr (ed): Principles of Nuclear Medicine. Philadelphia, Saunders, 1968, pp 599-620

chapter 3

THYROID

One sometimes wonders to what degree nuclear medicine would have progressed were it not for the ideal model the thyroid gland provides for studying an organ system with radionuclides. Certainly the initial use of radioactive materials for medical purposes was closely linked to the introduction of radioactive iodine for use in thyroid function studies, imaging, and treatment of hyperthyroidism. To this day no organ has been so directly associated with studies using radioactive materials as has the thyroid.

This chapter is concerned primarily with radionuclide studies specifically related to thyroid structure or function. There is no attempt to cover thyroid physiology and disease. For more extensive treatment of these subjects, the interested reader may consult the references at the end of this section.

Because the thyroid gland is involved with the trapping of inorganic iodide and its organification and incorporation into thyroid hormone, it may be readily understood that the same processes occur with tracer amounts of radioiodine. Thus practically all radionuclide studies of thyroid structure and function utilize one of the isotopes of iodine.

^{131}I-sodium iodide is the radionuclide most commonly employed today. Its relatively long half-life (8 days) and principal gamma emission of 363 keV make it convenient for practical handling purposes and for gamma detection. Recently ^{125}I is being used with increasing frequency. Although it has a half-life of 57 days, its low gamma energy (35 keV) sometimes makes detection difficult. An advantage of ^{125}I, which decays by internal conversion, is the absence of beta emission. Another isotope that shows great promise for studying the thyroid is ^{123}I. Its half-life of 13 hours and principal gamma emission of 159 keV make it an attractive material, but, because it is cyclotron-produced, its availability is dependent on the proximity of the nuclear medicine department to such a facility. During the past several years the use of ^{99m}Tc-sodium pertechnetate has also been increasing. Apparently the pertechnetate ion (TcO_4^-) is taken up in the thyroid in a manner similar to iodide but does not become organified. Unless otherwise stated, all reference to radionuclide studies of the thyroid in this chapter involve the use of ^{131}I as the tracer.

Following its ingestion and absorption from the intestinal tract inorganic iodide gains entry to the thyroid by two mechanisms: diffusion and trapping. The temporary increase in plasma iodide concentration allows the material to enter the gland by simple diffusion. Through trapping the thyroid is able to extract large quantities of iodide from the plasma, far in excess of what could be accomplished by diffusion alone. After this, the iodide undergoes oxidation and becomes organically bound to the tyrosine molecule (iodotyrosine). Two iodotyrosine molecules become coupled and enter into the formation of iodothyronines or thyroid hormone. Actually, thyroid hormone exists in the form of both tetraiodothyronine (T4) and triiodothyronine (T3).

The thyroid hormone is not present as free T3 or T4 in the gland but becomes bound to thyroglobulin (TG), the principal component of thyroid colloid; it is in this state that it is stored in the gland. Release of the hormone to the circulation, an action mediated by the pituitary through the secretion of thyroid-stimulating hormone (TSH), occurs as the TG undergoes hydrolysis by a proteolytic enzyme. The liberated thyroid hormone becomes attached to specific plasma proteins: thyroid-binding globulin (TBG) and thyroid-binding prealbumin (TBPA). These bonds are strong but reversible.

The concentration of thyroid hormone in the blood may be determined by measuring the level of iodine in the thyroid hormone bound to protein—known as protein-bound iodine (PBI). A small quantity of thyroid hormone (less than 1 percent) is not bound to these plasma proteins but circulates freely, making it possible to measure the free T4.

The steps by which inorganic iodide is ultimately incorporated into thyroid hormone are shown in Figure 1. Trapping (or transport) of inorganic iodide may be prevented at (1) by administering monovalent cations such as thiocyanate and perchlorate; pertechnetate also falls in this class. Agents like propylthiouracil act at (2) by interfering with the organic binding of iodine and the coupling reactions.

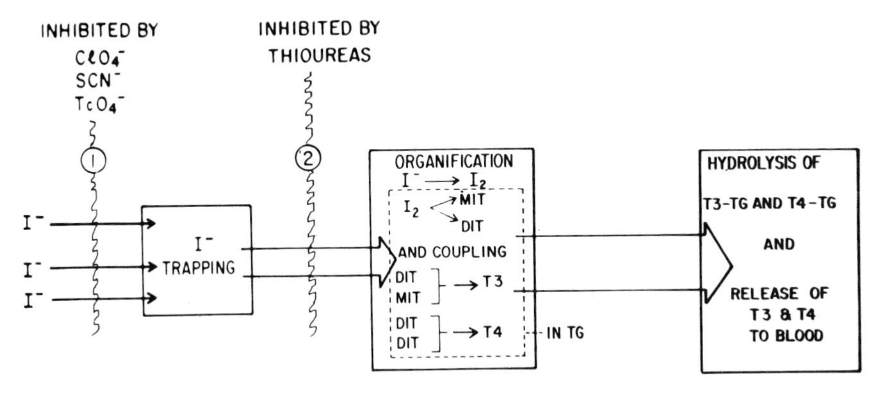

Fig. 1. Iodide trapping and organification. Trapping is prevented at (1) and organification blocked at (2). TG, thyroglobulin. MIT, monoiodotyrosine. DIT, diiodotyrosine.

RADIOACTIVE IODINE UPTAKE

Following administration and absorption, a tracer dose of ^{131}I enters the circulation and is cleared from the plasma by one of two avenues: excretion by the kidney or uptake into the thyroid.[1] The renal clearance rate of iodide is approximately twice that of the thyroid in the normal subject. The portion of the administered dose of ^{131}I that enters the thyroid is commonly known as the radioactive iodine (RAI) uptake.

The uptake of radioiodine into the thyroid gland usually reaches a peak about 24 hours after administration. Normally between 15 percent and 35 to 40 percent of the initial dose of radioactive material enters the thyroid by this time. Of course it is of great convenience to measure the RAI uptake at 24 hours, but values obtained 18 to 30 hours after ^{131}I administration are very close to the 24-hour figure, so the time of uptake determination need not be adhered to rigidly.

Technique

Because the radioactivity in the thyroid gland is determined through external monitoring, it is necessary that the ^{131}I-sodium iodide to be administered be put in a similar geometric setting to determine its radioactivity. The dose of radioactive material is placed in a recess in the neck phantom, usually made of Lucite, to simulate the location of the ^{131}I in the thyroid gland. Before giving the patient the radioiodine, its radioactivity in the neck phantom is determined with the space bar set so the detector is approximately 25 cm from the sample in the phantom. The patient then ingests the radioactive material, and the radioactivity in the patient's thyroid is measured 24 hours later in the same manner as that determined for the administered dose. After correcting the latter for 1 day's radioactive decay of ^{131}I (91 percent), the percentage uptake is calculated by comparing the radioactivity in the patient's thyroid to that in the dose of ^{131}I-sodium iodide:

$$\text{RAI uptake} = \frac{\text{cpm in patient's thyroid}}{\text{cpm in administered dose} \times 0.91} \times 100$$

Although most laboratories regard 15 percent to 35 or 40 percent as the normal range for the 24-hour RAI uptake, it is important that normal values be determined in each department. Only by performing this test on a large number of healthy subjects can a meaningful normal value for a particular locale be ascertained.

Increased RAI Uptake

Thyrotoxicosis. If excess hormone is being elaborated and discharged by the thyroid gland, as in hyperthyroid states, a greater than normal amount of radioiodine enters the gland. Thus the 24-hour RAI uptake is unusually elevated (greater than 35 percent).

In some cases of thyrotoxicosis iodine turnover is accelerated to the degree that by 24 hours after radioiodine administration much of the ^{131}I is incorporated into thyroid hormone and discharged from the gland into the circulation. The 24-hour RAI uptake values in these patients are in the normal range, which is simply a reflection of the rapid uptake and discharge of the ^{131}I. The values may thus be misinterpreted as indicating a euthyroid state.

To reckon with this situation, it is often helpful to determine RAI uptake in the hyperthyroid suspect at 3 to 6 hours following tracer administration as well as at 24 hours. In the normal subject the 6-hour RAI uptake is no greater than 25 percent, whereas a patient with thyrotoxicosis may have an uptake well above this value. However, it bears emphasis that it is uncommon for the 24-hour RAI uptake to be low in the hyperthyroid patient. Furthermore, in some hyperthyroid patients the RAI uptake may not be elevated at 6 or 24 hours. This is not infrequent, for example, in cases of single or multiple toxic adenomas.

Other Causes. Aside from hyperthyroidism, any situation characterized by an avidity of the thyroid gland for iodine is associated with an elevated 24-hour RAI uptake. Thus in cases of iodine starvation the thyroid attempts to trap as much of the iodide presented to it as it possibly can. This, of course, results in a relatively high RAI uptake.

Another related circumstance is seen with the iodine-rebound phenomenon. If, for example, a patient is treated with thyroid hormone, effectively reducing or shutting off TSH secretion through the feedback mechanism, sudden withdrawal of thyroid therapy results in an outpouring of TSH and so an increased iodide uptake by the thyroid gland. The elevated RAI uptake in such cases is transitory, continuing until there is resumption of normal thyroid hormone secretion. A rebound phenomenon is also seen in patients in whom antithyroid therapy (e.g., propylthiouracil) is suddenly withdrawn.

In some cases of nephrosis there is an unusual loss of protein in the urine including TBG and TBPA. The thyroid gland partially compensates for the reduction in thyroid hormone in the blood by increasing production of the hormone and increasing iodide trapping. These phenomena are reflected in an elevated RAI uptake.

Finally, some enzymatic defects are manifested by an elevated RAI uptake, as are certain specific thyroid disorders.

Reduced RAI Uptake

Hypothyroidism. Patients with hypothyroidism of course have a very low 24-hour RAI uptake—a reflection of reduced or absent thyroid hormone elaboration and discharge by the gland. It is of the utmost importance to distinguish between primary hypothyroidism and that associated with pituitary insufficiency (secondary hypothyroidism). In the latter, because a defect exists in the secretion of one or more tropic hormones by the pituitary gland, administration of exogenous TSH produces an elevated 24-hour uptake, often to about twice the initial value. In primary hypothyroidism TSH stimulation does not alter the uptake. It is common for the 24-hour RAI uptake in primary hypothyroidism to be close to zero. The uptake may be 10 to 15 percent in the secondary variety, probably due to some endogenous TSH secretion.

It is important to determine which of these two entities is present so as to initiate proper treatment. If thyroid replacement therapy is given to a patient with secondary hypothyroidism, the exogenous thyroid hormone may influence the feedback mechanism to the pituitary, causing a diminution or cessation in secretion of other pituitary tropic hormones and thus precipitating a possible adrenal crisis. On the other hand, thyroid hormone replacement is the treatment of choice in primary hypothyroidism.

Other Causes. The most common cause for factitious depression of the 24-hour RAI uptake is prior administration of iodinated compounds. This results in a greatly expanded amount of iodine available for uptake by the thyroid gland. Thus the opportunity for a tracer dose to be trapped by the thyroid is greatly reduced. In such a situation only a small amount of the administered radioiodine enters the gland, and the calculated 24-hour RAI uptake is spuriously low.

Of the oral medications, Lugol's solution is a frequent offender. Multiple vitamins in liquid form are often suspended in solutions containing iodine. Iodinated contrast materials used for roentgenologic studies are also associated with iodine contamination. Parenteral administration of contrast materials used for arteriography and intravenous urography can lead to false depression of the RAI uptake for about 6 weeks. Materials used for oral cholecystography undergo a continuous cycle of biliary excretion and reabsorption, resulting in reduced thyroidal uptake of RAI for approximately 6 months. Intrathecal administration of iodinated contrast material for myelography or cisternography may result in false depression of the RAI uptake for a number of years. Administration of thyroid hormone results in a lowering of the RAI uptake, probably by depressing the secretion of TSH by the pituitary.

OTHER THYROID FUNCTION STUDIES

The RAI uptake test is but one of many studies available to assist the physician in assessing the thyroid status of an individual. Tests to evaluate thyroid function may be grouped into two general categories: (1) studies related to the uptake and discharge of radioiodine by the thyroid gland, and (2) those concerned with determining the status of thyroid hormone in the blood, either by direct measurement or by inference. Several of the less frequently employed radionuclide thyroid studies are not included here, but interested readers may consult the references at the end of this chapter.

RAI Uptake and Discharge

The study in this group that has received the greatest emphasis is the RAI uptake—a testimonial to both the useful information it provides and the relative simplicity of the procedure. In addition to the RAI uptake, accumulation of ^{131}I in the thyroid may be studied through the *thyroid clearance* test. In this procedure the patient is given approximately 100 μCi ^{131}I-sodium iodide intravenously, and the clearance of the ^{131}I by the thyroid from the plasma is calculated. This is determined by comparing the radioactivity over the thyroid and that in the peripheral blood at intervals during the 30 minutes after radionuclide injection. The clearance is measured in terms of the volume of blood cleared by the thyroid (normally about 15 ml/minute). It is a difficult test to perform and is utilized almost exclusively when hyperthyroidism is thought to exist but when there have been inconclusive or conflicting laboratory findings. The thyroid clearance in the hyperthyroid subject is usually greater than 30 ml/minute.

The discharge of ^{131}I by the thyroid may also be utilized to assess thyroid function. This may be studied via the *$PB^{131}I$ test* and the *conversion ratio*—these two studies, along with the thyroid clearance, are used almost exclusively for detecting the hyperthyroid state.[2]

The $PB^{131}I$ is simply a reflection of the amount of administered ^{131}I (given for the purpose of RAI uptake studies) that has become incorporated and discharged in thyroid hormone. Because the $PB^{131}I$ plateaus 48 to 96 hours after tracer administration, it is convenient to measure it at 72 hours. The radioactivity of the $PB^{131}I$ per liter of plasma is determined at that time and compared with the radioactivity of the administered dose (per liter of plasma). This comparison is expressed in percentage form and in the hyperthyroid subject is greater than 0.27 percent.

A related study is the conversion ratio. This represents the proportion of ^{131}I, given for the RAI uptake, that has become incorporated into $PB^{131}I$ when compared to the total amount of radioactivity due to ^{131}I in the blood at a given time. The test is performed 24 hours after tracer administration; a conversion ratio greater than 50 percent at this time is consistent with hyperthyroidism.

Thyroid Hormone in Blood

During recent years in vitro studies of thyroid function have gained a wide measure of acceptance because of their accuracy and convenience (a simple blood sample is obtained from the patient). These studies are concerned with determining the status of thyroid hormone in the blood either by direct measurement or by inference.

The concentration of thyroid hormone in the blood may be determined in one of two ways: (1) indirectly by measuring the hormonal iodine content, as is done in the PBI; or (2) directly by measuring the blood level of thyroxine itself. Because thyroid hormone becomes bound to TBG and TBPA when it is discharged from the thyroid gland, the level of thyroid hormone in the blood may be estimated by measuring the iodine content in thyroid-bound proteins. These proteins are usually precipitated from serum, and the PBI so determined is usually between 4 and 8 $\mu g/100$ ml in the normal subject.

However, the PBI may become contaminated from both exogenous and endogenous sources. The ingestion of iodinated compounds in foodstuffs and medications (e.g., vitamin preparations) may falsely elevate the PBI. Of even greater concern is the false elevation caused by parenteral administration of contrast material used in roentgenologic studies. The PBI may also be increased in those situations in which iodinated proteins (e.g., mono- and diiodotyrosines) are secreted by the thyroid gland, as in Hashimoto's disease. Although these iodinated proteins do not represent thyroid hormone, they contribute to the protein-bound iodine value.

To some degree the PBI may be refined by employing the butanol-extractable iodine (BEI) test. Although both thyroid hormone and iodinated proteins are readily soluble in acid-butanol, the addition of Blau's reagent results in the extraction of only thyroid hormone. However, other iodinated materials (e.g., x-ray contrast media) are also butanol-extractable so the BEI is really of only limited usefulness.

Although until recently the PBI was considered by many thyroidologists the single most valuable test of thyroid function (and still is by some) the frequency of possible exogenous iodide contamination is a limiting factor. The problem of inaccurate results due to exogenous iodide is practically excluded when blood thyroxine levels are measured directly. Although several methods may be employed, this determination is best done by utilizing a competitive protein-binding analysis (Chapter 7).

Using ethanol, thyroxine is extracted from a sample of the patient's serum and added to a solution of TBG in which there is a known amount of labeled thyroxine. The added thyroxine displaces labeled thyroxine from the TBG; the radioactivity of the labeled thyroxine is determined; and the level of thyroxine in the patient's serum sample is then calculated.

False values of serum thyroxine determined by this method may be encountered when there is an unusual elevation or reduction of plasma proteins. If the proteins are increased (e.g., pregnancy, estrogen therapy), more thyroxine is extracted and the displacement of labeled thyroxine from the TBG is increased. The opposite is true in those cases in which the protein level is reduced (e.g., nephrosis, testosterone therapy). Because exogenous iodides do not influence T4 binding by TBG, administration of such materials is not a cause for concern.

The status of thyroid hormone in the blood may also be studied by inferential evidence obtained in the T3 resin uptake and free T4 tests. TBG has receptor sites for thyroid hormone that in the hyperthyroid patient are occupied to a high degree, while in the hypothyroid state relatively few are occupied. In the *T3 resin uptake* (Fig. 2) the patient's serum, containing TBG, is added to a known amount of radioactive T3 labeled with ^{125}I or ^{131}I. Using ^{131}I as the example, the ^{131}I-T3 becomes bound to the TBG at receptor sites that are still unoccupied. If an adsorbing material such as a resin sponge is placed in the test tube, the free ^{131}I-T3 that remains after all the receptor sites in the TBG are filled is adsorbed to the sponge.

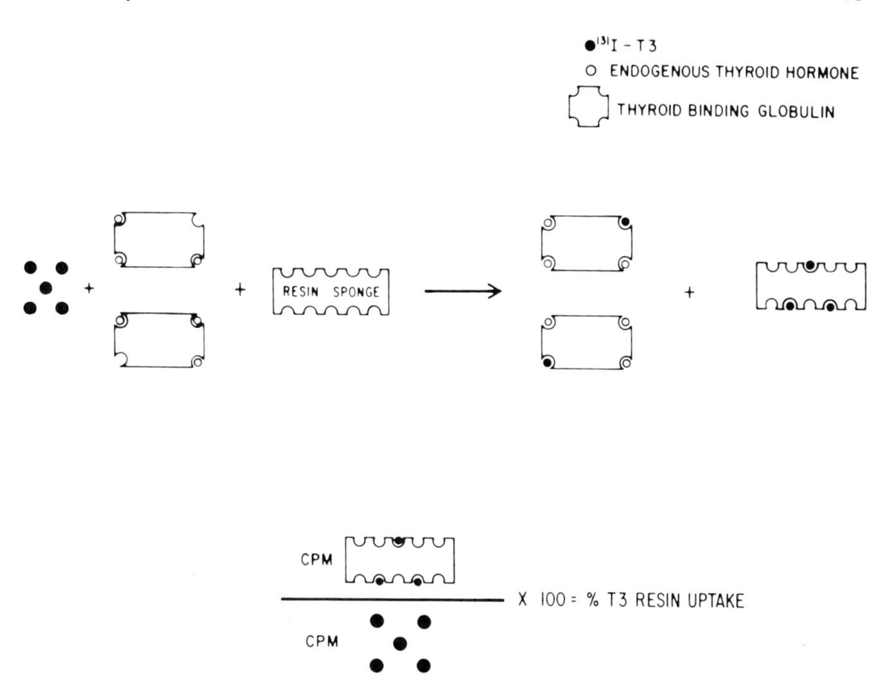

Fig. 2. T3 resin uptake. The hyperthyroid patient has few unfilled receptor sites in TBG so the excess ^{131}I-T3 becomes attached to the resin sponge, whereas in hypothyroidism there are numerous free receptor sites to which the ^{131}I-T3 can become attached. The T3 resin uptake is determined by comparing the radioactivity remaining in the resin sponge to the ^{131}I-T3 that was added to the system.

Because a relatively large number of TBG receptor sites are already occupied by endogenous thyroid hormone in the hyperthyroid patient, few receptor sites are available for the ^{131}I-T3 to enter and the quantity of ^{131}I-T3 that adheres to the resin sponge is relatively large. On the other hand, if few receptor sites are occupied by endogenous thyroid hormone, as in hypothyroidism, more ^{131}I-T3 can enter these unoccupied sites and less will be available for adherence to the resin sponge.

The radioactivity in the resin sponge is compared with the total amount of ^{131}I-T3 that was provided and is expressed as a percent. The normal range is usually 25 to 35 percent, with values greater than 35 percent being indicative of hyperthyroidism.

A great advantage of the T3 resin uptake is that it is not influenced by exogenous iodides. False T3 resin uptake results are sometimes associated with elevation or depression of plasma proteins. Providing there has not been an increase in endogenous thyroid hormone secretion, an increase in TBG due to a general elevation of plasma proteins makes available additional unoccupied thyroid hormone receptor sites in the TBG. Such a rise occurs with estrogen therapy or during pregnancy. More ^{131}I-T3 enters the additional unoccupied receptor sites, less is available for adsorption to the sponge, and the T3 resin uptake is low. Reduction in plasma proteins, as seen with testosterone therapy or in the proteinuria of nephrosis, results in fewer binding sites for endogenous thyroid hormone. As a result, the available unoccupied sites in TBG are reduced, more ^{131}I-T3 is adsorbed to the sponge, and the T3 resin uptake is increased.

During normal pregnancy the T3 resin uptake is reduced. A normal uptake in the pregnant patient may be indicative of hyperthyroidism.

The *free T4 test* has gained increased prominence in recent years. If a quantity of ^{131}I-T4 of known radioactivity is added to a patient's serum, the labeled T4 is distributed in much the same fashion as endogenous T4 after equilibrium is reached—i.e., the portion of ^{131}I-T4 bound to TBG and that which is free are proportional, respectively, to the portions of endogenous T4 that are bound and free. The unbound (or free) ^{131}I-T4 may be recovered by dialysis as it crosses a semipermeable membrane and its radioactivity then counted. By comparing the radioactivity of the free ^{131}I-T4 and the total amount used, the percentage of free T4 can be calculated. Normally about 0.05 percent circulates as free T4 and the remaining 99.95 percent is bound to TBG. The free serum thyroxine can then be determined by multiplying the percent of free thyroxine by the concentration of the thyroxine in the blood. Unlike the T3 resin uptake, the free T4 determination is influenced by iodide contamination of the serum.

Clinical Considerations

Any number of laboratory examinations can be utilized to investigate a patient for the possibility of thyroid disease. However, in screening patients who are clinically suspected of having hypo- or hyperthyroidism, it is useful to obtain one or more of the tests that determine the status of thyroid hormone in the blood as well as one of the radioiodine uptake studies.

Blood should be obtained for PBI, serum thyroxine, T3 resin uptake, or the free T4 test. If there has been no prior contamination with iodides, the PBI or free T4 may be used. In those cases in which iodine contamination is known or suspected, the serum thyroxine or T3 resin uptake is useful.

The RAI uptake is helpful in the evaluation of all patients with possible thyroid disorders. If the patient is considered to have hyperthyroidism, a 6-hour RAI uptake should be performed as well as the 24-hour study. The thyroid clearance, PB^{131}I, and conversion ratio should be reserved for those cases in which additional studies are needed to confirm the presence or absence of hyperthyroidism. Currently many new and valuable tests of thyroid function are being developed using competitive protein binding and radioimmunoassay techniques. (See also section on Hyperthyroidism.)

THYROID IMAGING

Because ^{131}I is ultimately incorporated into thyroid hormone, it follows that the presence of this radioactive material in the gland may be utilized for radionuclide imaging. It is customary to perform thyroid scans at approximately the same time as uptake studies are done, about 24 hours after administering the ^{131}I. Although smaller doses may be used if only the RAI uptake is to be determined, 50 μCi ^{131}I-sodium iodide is usually sufficient for thyroid scanning if the uptake is believed to be in the normal range.

The thyroid scan is to some degree a manifestation of thyroid function insofar as radioiodine uptake is concerned. The gland is imaged with better definition when the uptake is high than when it is very low. Peripheral areas where the thyroid mass is much less than the central mass may not be seen at all in the patient whose RAI uptake is low, whereas these same regions may be well visualized in the patient with an elevated uptake.

In a normal thyroid scan (Fig. 3) each lobe usually measures no more than 4.5 to 5.0 cm longitudinally, but this is not precise and thyroid size varies with body build. The suprasternal notch should always be marked as a reference point on a thyroid scan in case there is substernal extension of the gland.

Variants of normal are not infrequent. A common finding is an asymmetrical gland in which one lobe is higher than the other (Fig. 4), and a pyramidal lobe may appear in close relation to an upper pole (Fig. 5A). Sometimes a lingual thyroid or a remnant of the thyroglossal duct containing

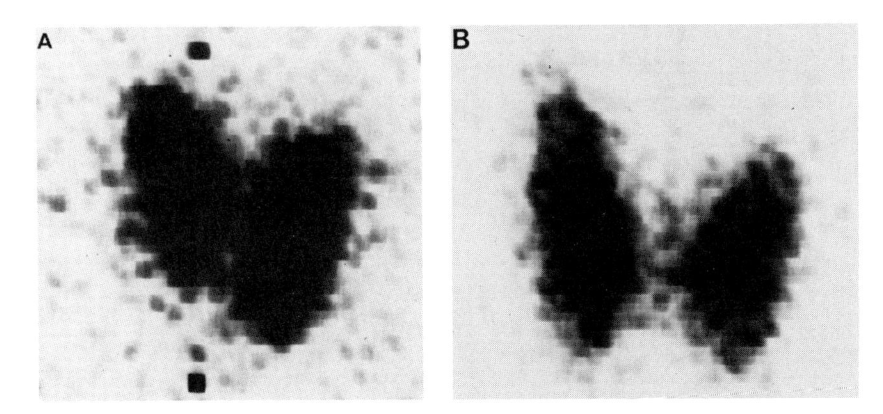

Fig. 3. Normal thyroid scans in which both lobes are similar in size.

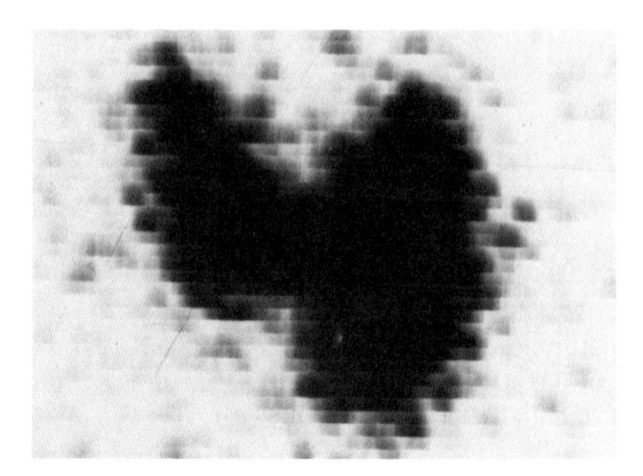

Fig. 4. Asymmetry of the thyroid gland with relative elevation of the right lobe.

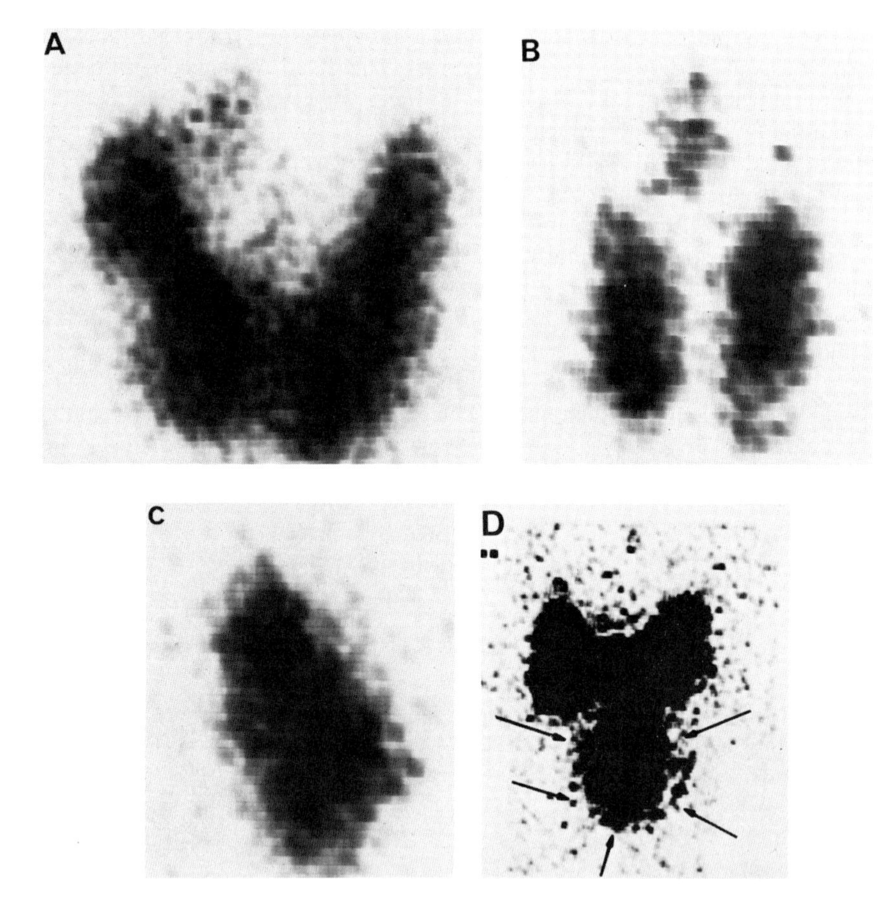

Fig. 5. Accessory lobes. A. Pyramidal lobe closely related to the right upper pole. B. Lingual thyroid. C. Congenital absence of the left lobe. D. Substernal extension, indicated by arrows.

thyroid tissue is seen on the scan (Fig. 5B), and rarely there is congenital absence of one lobe (Fig. 5C). Occasionally, substernal extension may be imaged (Fig. 5 D).

Thyroid Nodules

Certainly one of the cardinal uses of thyroid scanning is to identify and evaluate thyroid nodules.[3] Often from physical examination it can be determined that there is a single nodule or that the gland is multinodular. Scanning helps determine if the nodule is functioning or nonfunctioning.

It is essential to correlate the physical and scan findings in assessing thyroid nodules. Although this may also be accomplished using a scintillation camera, a rectilinear scanner is less cumbersome and more precise. One method to achieve this correlation is as follows. With the patient in the hyperextended scanning position, the nodule(s) is located by physical examination and the periphery (of each) is marked with three or four points. The light localizer in the detector is placed over each point marked on the patient's skin, and the corresponding point under the printing stylus is marked on the dot scan. Thus the coinciding areas of interest can easily be studied. The marking can be performed before the scan is started or at the completion of the study.

Perhaps the most vexing problem confronting the physician in the area of thyroid disease is how to manage the patient with a nodule. Bearing in mind that there may be exceptions, the following is generally true. Suspicion of malignancy is aroused when a thyroid nodule is solitary and presents on the scan as an area of nonfunction. Nevertheless, the possibility of malignancy in a solitary, nonfunctioning nodule is extremely low, probably no greater than 4 percent. Although nonfunctioning nodules are nine times more common in females than in males, there is a much higher proportion of malignancy in these nodules in men. With advancing age, the occurrence of benign goiter increases. For both males and females the highest incidence of thyroid malignancy is during the third and fourth decades. In addition, neoplastic involvement is a prime consideration in the patient with a thyroid nodule who gives a history of having received radiation to the neck during childhood. On the other hand, the incidence of malignancy in a hyperfunctioning nodule is practically zero. For purposes of scanning, it is convenient to classify thyroid nodules as follows:

- Nonfunctioning nodules
  - Solitary
  - Multiple
- Functioning nodules
  - Nodules with functioning thyroid tissue
  - Autonomously functioning nodules

Solitary Nonfunctioning Nodules. Nonfunctioning ("cold") nodules are seen on scans as discrete areas of non- or hypofunction. The latter are sometimes difficult to delineate through scanning.

The scan of a patient with a solitary nodule at the right lower pole of the thyroid is shown in Figure 6. There is a somewhat rounded area of absent radioactivity underlying the palpable nodule, *suggesting* the presence of a nonfunctioning nodule. This abnormality was clarified with a thyroid stimulation study. The patient received 10 units TSH intramuscularly for 3 days and on the third day was also given an oral dose of 50 μCi ^{131}I-sodium iodide. The scan performed the next day shows persistence of the area of absent radioactivity, clearly defining a truly nonfunctioning nodule. On the basis of the physical examination findings (the solitary nodule) and the demonstration by scan that it was nonfunctioning, surgery was performed; the patient had an anaplastic carcinoma.

Persistence of the nonfunctioning nodule following TSH stimulation certainly suggests the possibility (though statistically remote) of neoplasm. It certainly indicates an absence of thyroid tissue that could be stimulated to take up ^{131}I in the area of apparent nonfunction. There are countless other disorders that may show similar characteristics on physical examination and scanning (e.g., cyst, hemorrhage, benign adenoma).

The scan of a patient with a left lower pole nodule is shown in Figure 7 before and after TSH stimulation. Although there was persistence of the nonfunctioning area, at surgery this was found to be a benign adenoma.

Another patient was found to have a solitary nodule in the right lobe of the thyroid gland (Fig. 8). There is a somewhat irregular area of absent

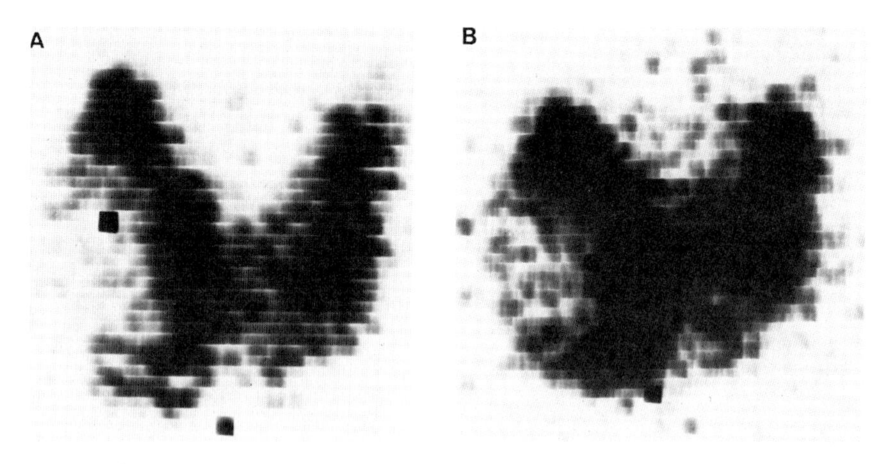

Fig. 6. A. Scan of thyroid showing evidence of nonfunctioning nodule at right lower pole. B. Scan of same patient after TSH stimulation; note persistence of the nonfunctioning nodule. Anaplastic carcinoma of the thyroid was discovered at surgery.

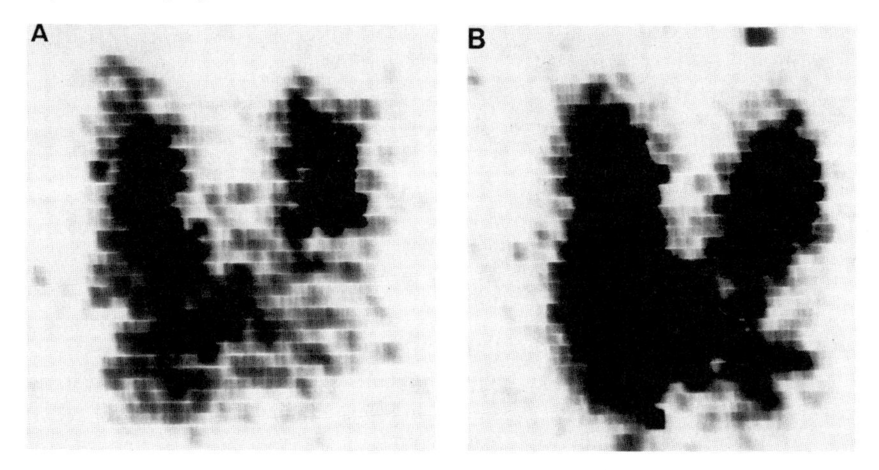

Fig. 7. A. Reduced radioactivity at left lower pole in region underlying palpable nodule. B. Scan of same patient following TSH stimulation showing persistence of the nonfunctioning nodule. Surgery disclosed a benign adenoma.

radioactivity underlying the palpable thyroid nodule that suggests the presence of a nonfunctioning nodule. The scan performed after TSH stimulation shows only minor reduction in radioactivity, and there is no longer visualization of a relatively large nonfunctioning nodule. The patient later gave a history of having had an episode of subacute thyroiditis 3 years before.

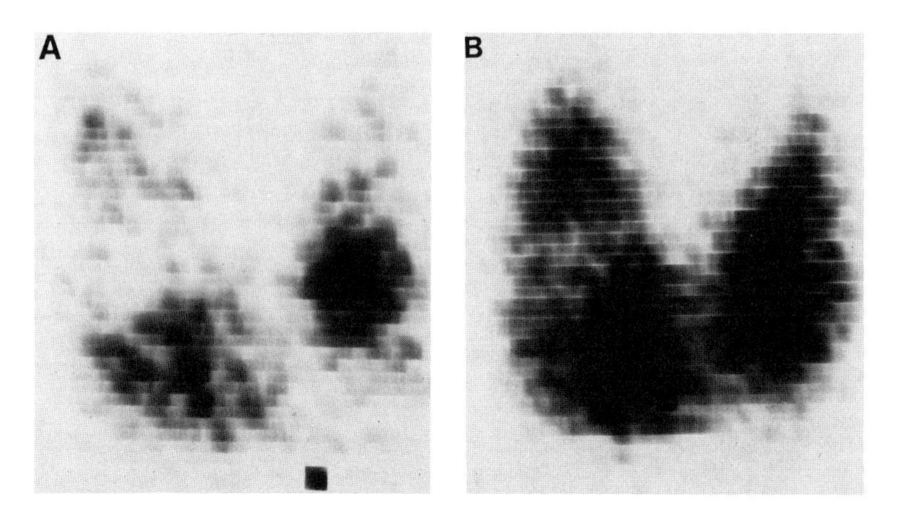

Fig. 8. A. Nonfunctioning nodule in right lobe. B. After TSH stimulation the area of nonfunction is no longer sharply defined. Patient had previous episode of subacute thyroiditis.

The palpable nodule simply represents a fibrotic area associated with the previous thyroiditis. After TSH stimulation, ^{131}I was able to enter the surrounding, relatively hypofunctioning thyroid tissue. The remaining small area of absent radioactivity is most likely associated with localized fibrous tissue replacement.

Multiple Nonfunctioning Nodules. Multiple nonfunctioning nodules are most commonly encountered in patients with multinodular nontoxic goiters. The patient usually gives a history of having had a goiter for many years and often once resided in an area considered a "goiter belt" or with other forms of iodine deprivation. This entity is illustrated in Figure 9, where numerous colloid cysts are seen throughout the enlarged gland.

A less frequent and extremely rare cause of multiple nonfunctioning nodules is neoplasm. Such patients usually present with enlarging, firm thyroid nodules and no history of goiter.

Nodules with Functioning Thyroid Tissue. Thyroid nodules most commonly occur in association with thyroid tissue that is functioning and has un-

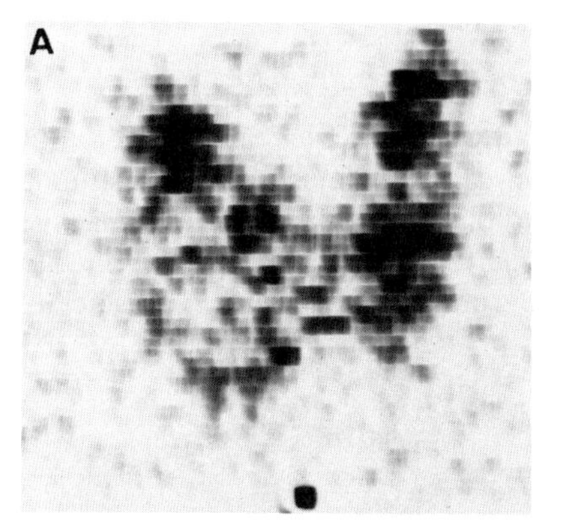

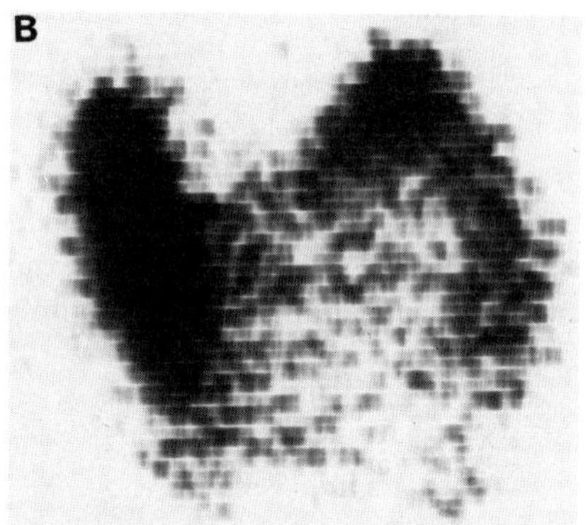

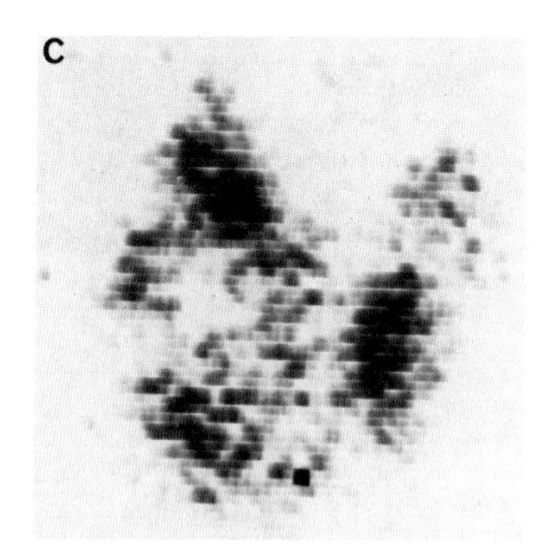

Fig. 9. Scans of patients with multinodular goiter. Note the large, somewhat rounded areas of absent radioactivity.

dergone some degree of hyperplasia. These nodules may be single or multiple. In many cases the nodule cannot be delineated on the scan, which shows no abnormalities. In others the scan shows slightly increased radioactivity resulting from the increased thyroid tissue mass connected with the palpable nodule. Sometimes an entire lobe is involved.

Occasionally, such nodules regress after thyroid therapy, presumably as a result of shutting off the TSH thought to be responsible for the localized stimulation to hyperplasia. However, visible and palpable results of such regression often takes many months and sometimes years.

Autonomously Functioning Nodules. If a thyroid nodule is functioning autonomously, it elaborates thyroid hormone completely independent of TSH. Through the feedback mechanism the secretion of thyroid hormone leads to a reduction or shutting off of TSH by the pituitary gland. As a result, little or no TSH is available to stimulate the portion of the gland outside the nodule. The autonomously functioning nodule is *suppressing* the balance of the gland.

A typical case is that of a 45-year-old female with a palpable left-sided thyroid nodule; she was otherwise asymptomatic. The 24-hour RAI uptake was 27 percent, and she was considered to be euthyroid. The scan in Figure 10A shows somewhat rounded, localized areas of increased radioactivity in the regions of the right upper and left lower lobes. Very little radioactivity is seen elsewhere in the gland. The areas of reduced or absent radioactivity indicate the portion of the gland that is suppressed.

Two confirmatory studies may be performed to establish whether the nodule is truly autonomous: a TSH stimulation study and thyroid suppression test. The scan performed following TSH stimulation (Fig. 10B) demonstrates that the portions of the thyroid previously suppressed are able to take up the ^{131}I when exogenous TSH is administered. The autonomously functioning nodules are not clearly defined on this scan. The scan of a patient with a single autonomously functioning right upper pole nodule is seen in Figure 10C. After suppression with T3 for 8 days the nodule continues to be clearly identified, but there is poor visualization of the other portions of the gland. This in turn demonstrates that the nodule is truly autonomous and cannot itself be suppressed by exogenous thyroid hormone.

The cases just illustrated are those of nontoxic autonomously functioning nodules. Such patients should be closely followed because thyroid hormone secretion may be increased to the point that thyrotoxicosis ensues. When associated with hyperthyroidism, the situation is that of a toxic autonomously functioning nodule. The developing hyperthyroid state is sometimes difficult to detect.

A 45-year-old man was hospitalized because of atrial fibrillation, nervousness, and a right-sided thyroid nodule. The 24-hour RAI uptake was 48 percent, and the scan suggested that the palpable nodule was autonomously

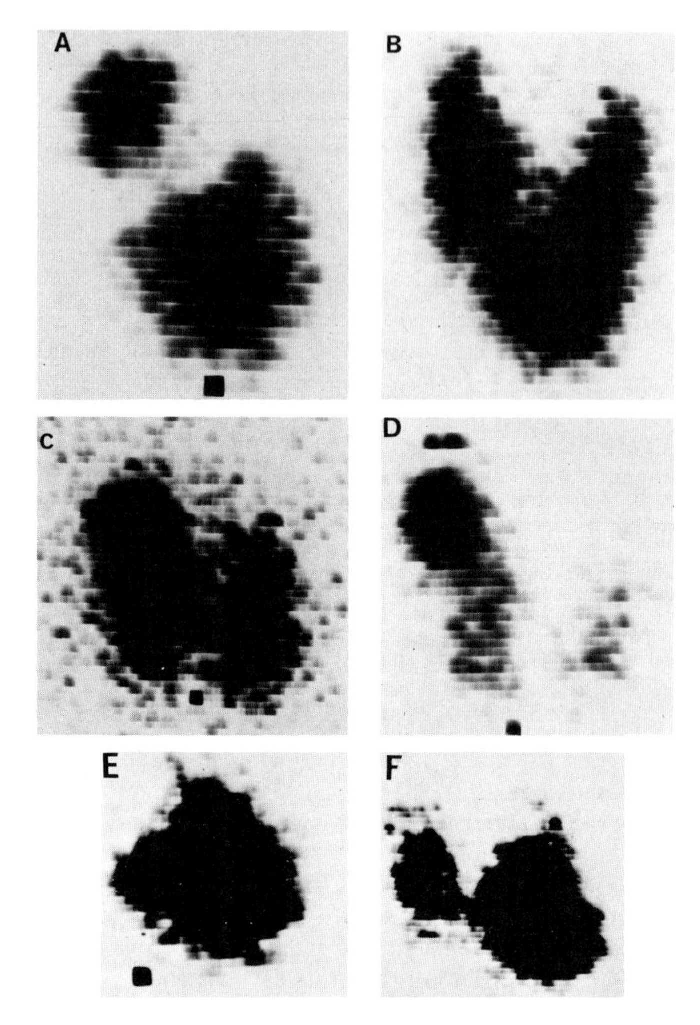

Fig. 10. Thyroid gland with two nontoxic, autonomously functioning nodules A. Before TSH stimulation. B. After TSH stimulation. C. Thyroid gland with single, autonomously functioning nodule. D. Same gland following suppression with T3. Thyroid gland with single, left-sided autonomously functioning nodule. E. Before TSH stimulation. F. After TSH stimulation.

functioning and suppressing the balance of the gland (Fig. 11). The PBI and T3 resin uptake were both in the hyperthyroid range.

The 24-hour uptake is sometimes deceptively low due in part to the small geographic area of hyperfunction; thus the opportunity for ^{131}I entering the nodule is much less, for example, than that encountered with the uniformly

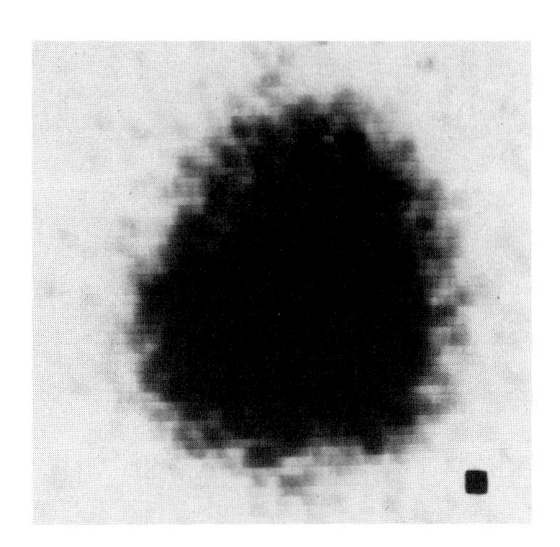

Fig. 11. Toxic autonomously functioning nodule.

increased uptake of Graves' disease. However, the PBI and T3 resin uptake are usually elevated.

The onset of hyperthyroidism in the patient with an autonomously functioning nodule is sometimes insidious and the manifestations often few. There is no dramatic development of the many signs and symptoms so commonly seen with the thyrotoxicosis of Graves' disease. Instead, there may be only a cardiac arrhythmia, weight loss, or nervousness associated with the palpable nodule. The toxic autonomously functioning nodule may be the masquerader of other disease entities.

Treatment is either surgical excision of the toxic nodule or radioiodine application. Because the uptake of ^{131}I by the gland is not uniform, the dose of radioactive material can not be calculated as is done for Graves' disease. Depending on the severity of the hyperthyroidism, an empirical dose of ^{131}I-sodium iodide (6 to 12 mCi) is commonly given. Medical therapy with such agents as propylthiouracil is usually unrewarding in the treatment of toxic autonomously functioning nodules.

Scintillation Camera Imaging

A great advantage of imaging the thyroid gland with a rectilinear scanner is that important physical findings, such as thyroid nodules, can be outlined directly on the scan itself. This is not done so easily when using a scintillation camera.

However, images of the thyroid obtained with a gamma camera using a pinhole collimator are usually of excellent resolution (Fig. 12).

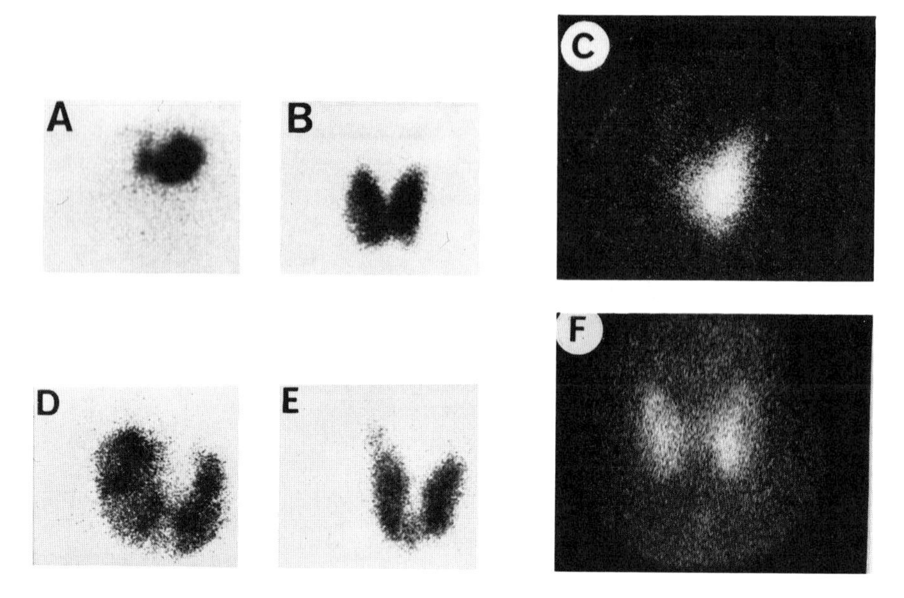

Fig. 12. Scintillation camera images of the thyroid using a pinhole collimator. A. Left-sided autonomously functioning nodule. B. Non-functioning nodule, right lower pole. C. Surgical absence of right lobe. D. Toxic, multinodular goiter. E. Right pyramidal lobe. (A-E. All with ^{131}I.) F. Normal thyroid imaging using ^{99m}Tc-sodium pertechnetate.

HYPERTHYROIDISM

While it is true that the employment of radionuclides for thyroid function studies and imaging play a prominent role in evaluating the hyperthyroid suspect, the preeminence of the clinican in the management of these patients should never be minimized. There is often a tendency to rely too heavily on nuclear medicine procedures when studying such patients.

Hyperthyroidism may be associated with one of three entities: (1) Graves' disease, (2) toxic multinodular goiter, and (3) toxic autonomously functioning nodule. By far the most frequent manifestation of hyperthyroidism appears with Graves' disease, which is characterized by goiter, thyrotoxicosis, and exophthalmos. This disorder occurs more frequently in females than males (by a ratio of nine to one) and most often during the third decade. On physical examination the gland is usually smooth and diffusely enlarged. Note the absence of nodularity on the scan in Figure 13.

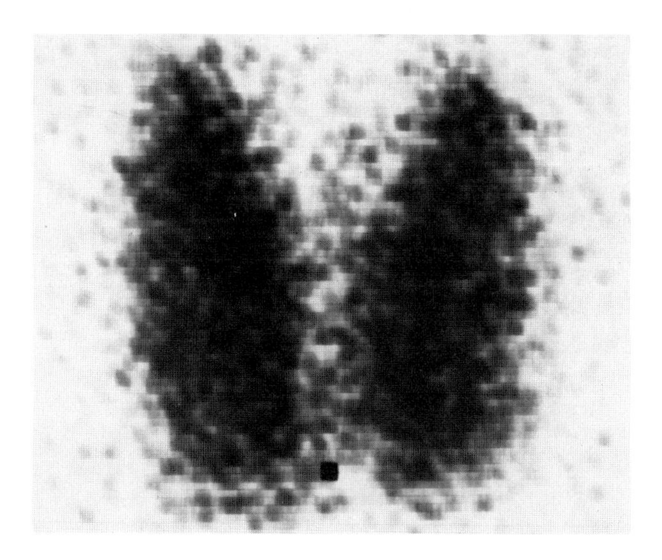

Fig. 13. Scan of patient with Graves' disease with elevated RAI uptake. Note the absence of nodularity.

In addition to the characteristic signs already mentioned, patients with Graves' disease usually have a multiplicity of symptoms; these may include extreme nervousness and insomnia, weight loss, sweating and heat intolerance, diarrhea, and cardiovascular manifestations such as palpitations and atrial fibrillation. The PBI, serum thyroxine, T3 resin uptake, free T4, and RAI uptake studies are all usually elevated in these patients.

There is now a large body of evidence suggesting that Graves' disease may be an autoimmune disorder, and that there is a strong familial association. A substance known as long-acting thyroid stimulator (LATS) has been found in the blood of many patients with Graves' disease. This material may be associated with some of the manifestations of the disorder, such as exophthalmos and pretibial myxedema.

The scan of a patient with toxic multinodular goiter is shown in Figure 14. Multiple areas of relatively increased radioactivity representing nodular formation are situated throughout the enlarged gland. Interspersed are areas of normally functioning thyroid tissue. This disorder often occurs as a consequence of nontoxic multinodular goiter. On physical examination the gland is reminiscent of what is seen on the scan and is enlarged and nodular.

The clinical manifestations of the toxic multinodular goiter are not nearly so obvious or dramatic as those seen in Graves' disease. The RAI uptake may be elevated or even in the normal range (see the section dealing with the thyroid suppression test).

The thyroid clearance, $PB^{131}I$, and conversion ratio are usually in the

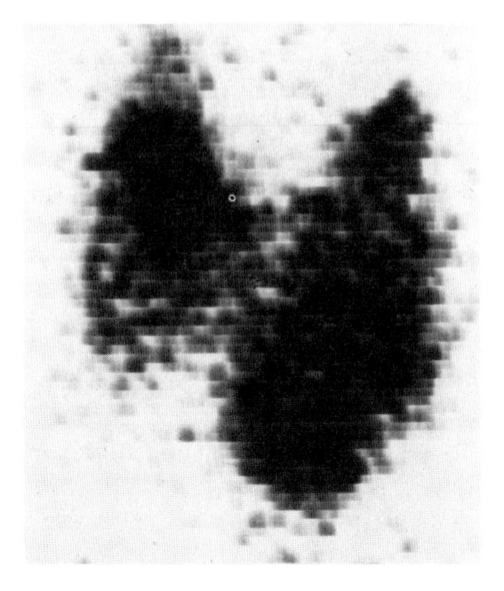

Fig. 14. Toxic multinodular goiter.

hyperthyroid range as are the blood studies of thyroid hormones. The only overt signs and symptoms, aside from thyromegaly, are associated with the cardiovascular system; these may include cardiac arrhythmias and palpitations.

The toxic autonomously functioning nodules, already discussed, may be single or multiple, and in each case there is suppression of ^{131}I uptake in the uninvolved portion of the gland. Symptoms and signs of hyperthyroidism in toxic autonomously functioning nodules are sometimes very few. As in patients with toxic multinodular goiters, the clinical picture is often dominated by cardiovascular findings (e.g., atrial fibrillation and tachycardia).

Further clarification of whether the nodule is autonomously functioning may be gleaned from thyroid suppression. The physician should be mindful of the hazard of administering TSH to the already hyperthyroid patient.

Clinical Considerations

The place of nuclear medicine procedures in the approach to the hyperthyroid suspect merits special consideration. Before any radioactive materials are administered the physician should anticipate one or more of the blood studies that should be performed, usually including PBI, serum thyroxine, T3 resin uptake, and free T4. The patient may then be given an oral dose of radioiodine for the RAI uptake and scan, the former to be performed at both 6 and 24 hours.

If hyperthyroidism is a strong possibility and the RAI uptake is not elevated, or if the RAI uptake is elevated and there is doubt as to the presence of hyperthyroidism, a thyroid suppression test (Werner test) may be performed.[4] The patient receives an oral dose of 75 μg T3 daily for 8 days and then another dose of ^{131}I for RAI uptake studies. In the absence of hyperthyroidism the exogenous T3 reduces or shuts off the outpouring of TSH by the pituitary gland through the feedback mechanism, and the uptake of ^{131}I into the thyroid gland is suppressed. Thus the RAI uptake at 6 or 24 hours is decreased by at least 20 percent in the nonhyperthyroid individual when compared to the RAI uptake done before suppression. On the other hand, if hyperthyroidism is present the uptake is relatively unchanged—i.e., suppression of ^{131}I uptake into the thyroid gland is absent because TSH secretion has already been shut off or reduced by secretion of thyroid hormone from the already overactive gland.

Several tests related to the uptake and discharge of radioactive iodine by the thyroid may be extremely helpful in arriving at a decision as to whether the patient has hyperthyroidism. A thyroid clearance of greater than 30 ml of plasma per minute certainly helps establish a diagnosis of hyperthyroidism. A PB^{131}I greater than 0.27 percent and a conversion ratio of 50 percent or more further confirm the presence of thyrotoxicosis.

Once the diagnosis of hyperthyroidism is made, the next consideration is the mode of therapy. Treatment falls into one of three broad categories: medical, surgical, and radioiodine application. The physician and patient together must decide what treatment program is to be followed. Medical therapy with such agents as propylthiouracil is usually considered the least complicated insofar as permanent sequelae and having leeway for treatment alterations are concerned. Anxious or uncooperative patients often find it difficult to adhere to the rigorous dosage schedules and prolonged periods of administration. Moreoever, the hazards of agranulocytosis cannot be ignored.

The problems of surgery are well known, and such postoperative complications as recurrent laryngeal nerve paralysis, hypoparathyroidism, and hypothyroidism must be considered. Nevertheless the results are rapid, and the method is sometimes preferred for cosmetic reasons and in young adults. Surgery may be the procedure of choice for the patient with a toxic autonomously functioning nodule.

Radioiodine Therapy

With the passage of time, restrictions in the use of radioactive iodine for the treatment of hyperthyroidism have become fewer and fewer. Of course radioactive materials of any kind should not be administered, if possible, to the pregnant or lactating female or to children and young adults. It may be generally stated that women of childbearing age who receive a therapeutic dose of ^{131}I for hyperthyroidism should delay conception for at least 1 year. Otherwise, there are no specific contraindications to the use of ^{131}I therapy in thyrotoxicosis because of possible harmful radiation effects.

During the early days of radioiodine treatment for thyroid disease, extremely large doses of radioactive iodine were given and the appearance of myxedema 2 to 3 years later was not uncommon. Since that experience the dosage schedule has been scaled down considerably and is now done with apparently improved methods.

Determination of a proper dose of ^{131}I in a patient with Graves' disease is undertaken with the objective of delivering 8,000 to 12,000 rads to the thyroid gland. Based on the method of Silver,[2] the dose is calculated on the basis of the gland's weight and the 24-hour RAI uptake.

Estimating the weight of the gland is open to the inaccuracies inherent in any subjective determination. However, with experience, approximations within 10 percent of the actual weight are attained by some examiners. The late Dr. Solomon Berson taught that if by palpation it were estimated that each thyroid lobe fits in a standard 1-ounce whiskey glass, the gland weighs 60 g.

According to Silver's method, for the first 50 g of thyroid tissue the amount of ^{131}I to be delivered to the gland is 80 μCi/g. For that part of the gland that exceeds 50 mg, 40 μCi ^{131}I is delivered for each gram of thyroid tissue. Take note of the term *delivered* for this is based on a 24-hour RAI uptake of 100 percent. Thus the general equation for the ^{131}I dose for the first 50 mg of thyroid tissue is:

$$^{131}\mathrm{I}\ (\mu\mathrm{Ci}) = \frac{50\ \mathrm{g} \times 80\ \mu\mathrm{Ci/g}}{\text{24-hour RAI uptake}}$$

The additional dose for anything greater than 50 g is given by the equation:

$$^{131}\mathrm{I}\ (\mu\mathrm{Ci}) = \frac{\text{weight over 50 g} \times 40\ \mu\mathrm{Ci/g}}{\text{24-hour RAI uptake}}$$

Example: Calculate the therapeutic dose of ^{131}I-sodium iodide to be administered to a patient with Graves' disease, whose 24-hour RAI uptake is 70 percent and whose gland is estimated to weight 65 g.

$$^{131}\mathrm{I}\ (\mu\mathrm{Ci}) = \frac{50\ \mathrm{g} \times 80\ \mu\mathrm{Ci/g}}{0.70} = 5{,}714\ \mu\mathrm{Ci}\ \text{or 5.7 mCi for first 50 g}$$

$$^{131}\text{I}\ (\mu\text{Ci}) = \frac{15\ \text{g} \times 40\ \mu\text{Ci/g}}{0.70} = 857\ \mu\text{Ci or } 0.9\ \text{mCi for remaining } 15\ \text{g}$$

Total ^{131}I dose = 5.7 mCi + 0.9 mCi = 6.6 mCi

The patient often experiences some subjective improvement within 3 to 4 days of radioiodine therapy, with marked relief noted by 3 to 4 weeks. In cases of severe thyrotoxicosis, radiation thyroiditis may ensue as a consequence of radioiodine therapy. This extremely rare and unusual complication takes place about 2 weeks after the dose of ^{131}I is given and results from a sudden discharge of thyroid hormone. It may sometimes be anticipated in the person with flagrant hyperthyroidism. To help prevent it the patient is treated for a week with propylthiouracil, this therapy being discontinued 3 to 4 days prior to ^{131}I treatment. Then 3 or 4 days after the radioiodine is administered, the propylthiouracil is resumed for 2 weeks. This regimen diminishes the large amount of thyroid hormone secretion that would otherwise occur.

Another, more frequently encountered complication is the accentuation of ophthalmopathy following RAI therapy. This may in part be related to the increased amounts of TSH secreted by the pituitary following therapy and the presence of LATS. Whatever the source of the problem, it may become extremely severe with conjunctivitis, corneal ulceration, and worsening of exophthalmos. Topical steroids are of benefit sometimes, and in extreme cases exophthalmos may be corrected by surgical means for cosmetic reasons.

Different problems exist in the therapeutic approach with ^{131}I for the patient with a toxic multinodular goiter or a toxic autonomously functioning nodule. Calculating the dose of radioiodine from the RAI uptake and weight of the gland is not so easily achieved.

Patients with toxic multinodular goiter generally require a larger dose than would be given to a person with Graves' disease in whom the mass of the thyroid gland is comparable. The 24-hour RAI uptake is frequently not elevated, so a smaller percentage of ^{131}I is delivered to the gland. In addition, the thyroid is not uniformly involved and radioiodine therapy to the uninvolved portions is not an objective of therapy. Some thyroidologists empirically treat the patient who has toxic multinodular goiter with 6 to 10 mCi ^{131}I, depending on the severity of the thyrotoxicosis.

Radioiodine therapy of the patient with a toxic autonomously functioning nodule was already discussed.

Follow-up. Patients who receive ^{131}I therapy for hyperthyroidism of any variety should be followed closely, especially during the first year, for evidence of recurrence or induction of hypothyroidism. Some objective as well as subjective indication of partial or complete response to the radioiodine treatment may be gleaned as early as 4 to 6 weeks after therapy. By this time the parameters used to estimate thyroid function (e.g., T3 resin uptake or serum thyroxine) should be in the normal range. If there has been no subjective improvement or if the relief of hyperthyroid symptoms has been only transient, and the aforementioned laboratory values are in the hyperthyroid range, then the hyperthyroidism has probably recurred or the patient was inadequately treated. This is best treated with a small dose of ^{131}I (2 to 3 mCi). The 24-hour RAI uptake can not be employed for many months because of the presence of glandular and circulating ^{131}I associated with the therapeutic dose.

Follow-up by clinical and laboratory investigation should be continued at intervals of 3 months during the year following radioiodine treatment to possibly detect recurrence of the hyperthyroid state or induction of myxedema. After that, it is incumbent on the nuclear medicine physician to help the patient make arrangements for continued follow-up by a physician.

THYROIDITIS

Patients with thyroiditis, both subacute and chronic, often present perplexing diagnostic problems. Subacute thyroiditis (De Quervain's disease) usually follows an upper respiratory infection and is characterized clinically by an enlarged thyroid or nodular formation with diffuse or localized tenderness. Because of follicular destruction and replacement by inflammatory tissue, areas of absent radioactivity are seen on the scan that can not, without a history of subacute thyroiditis, be distinguished from neoplasm or other disorders associated with a nonfunctioning area (Fig. 15). An additional effect of follicular damage is the relative inability of iodide to enter the gland, resulting in a very low 24-hour RAI uptake. On the other hand, iodinated tyrosines discharged from the gland lead to an elevated PBI. The BEI, however, is much lower.

Because of the exquisite tenderness at the site of nodular formation, subacute thyroiditis must be distinguished from hemorrhage into a nodule. Localized areas of follicular destruction may be reflected on the thyroid scan for years to come as sites of absent radioactivity.

Chronic thyroiditis (Hashimoto's disease) is an autoimmune disorder that is insidious in onset and characterized clinically by a gradually enlarging thyroid gland. The goiter is smooth and nontender. Although iodide becomes trapped in the gland, there is a defect in organification of iodine, which may be demonstrated by the perchlorate discharge test.[1] For this study the patient

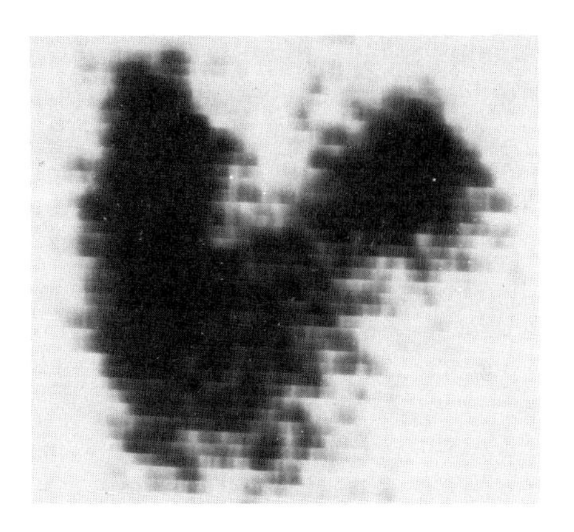

Fig. 15. Large area of reduced radioactivity in the left lobe associated with subacute thyroiditis.

is given approximately 200 μCi radioiodine by mouth, and the radioactivity is determined over the thyroid gland at 15-minute intervals for the next 2 hours. A dose of 1g potassium perchlorate is then administered by mouth. Because this material blocks iodide trapping, further iodide accumulation is prevented in the normal subject; in those with defects in organic binding the iodide already accumulated is discharged by the gland. If the radioactivity over the thyroid is again measured 30 minutes after perchlorate administration, there is a marked reduction in radioactivity because of ^{131}I discharge from the gland. On the other hand, the normal subject has thyroidal activity similar to that seen before the perchlorate was given, simply reflecting the failure to accumulate more ^{131}I after the blocking dose was given.

More specifically, patients with Hashimoto's thyroiditis have thyroid autoantibodies in their serum, and discovery of an elevated titer of these antibodies may establish the diagnosis. As in subacute thyroiditis iodinated proteins are secreted by the gland, resulting in a high PBI but very much lower BEI. Finally, the scan does not usually show the areas of absent radioactivity so commonly seen in subacute thyroiditis.

CARCINOMA OF THE THYROID

Carcinoma of the thyroid is associated with areas of absent radioactivity on scanning. The major types of thyroid neoplasm, in ascending order of malignancy, are papillary carcinoma, follicular carcinoma, and anaplastic carcinoma. Another type, medullary carcinoma, is frequently associated with pheochromocytoma.

In cases of follicular carcinoma in which there is metastatic involvement, approximately 10 percent have metastases that contain functioning thyroid follicles. Since such metastases can take up radioactive iodine, it is fruitful to search for them so they can receive the benefit of radioiodine therapy. In some patients with papillary carcinoma of the thyroid, metastases may also exhibit follicular structures capable of taking up radioactive iodine. If such metastases are thought to be present, no matter how remote the possibility, a thyroid metastatic survey should be performed. If a portion of the thyroid gland still shows functioning tissue, prior ablation with radioiodine enables functioning metastases to be more easily visualized by eliminating thyroidal ^{131}I uptake.

The patient is given 10 units TSH intramuscularly daily for 3 days and on the third day is also given a dose of 1.5 to 2.0 mCi ^{131}I-sodium iodide by mouth. Scanning is performed over the suspected areas 24 hours later. If this is unrewarding, the scanning should be repeated at 48 and 72 hours following radionuclide administration. If a metastasis does indeed show increased radioactivity and can be imaged, this is evidence of a functioning follicular structure that takes up radioiodine. Treatment is variable and may range from 30 to 100 mCi.

REFERENCES

1. Ingbar SH, Woeber KA: The thyroid gland. In Williams RH (ed): Textbook of Endocrinology, 4th ed. Philadelphia, Saunders, 1968, pp 105-286
2. Silver S: Radioactive Nuclides in Medicine and Biology, 3rd ed. Philadelphia, Lea & Febiger, 1968, pp 1-234
3. Green W, Senturia H, Packman R, et al: Management of the thyroid nodule. JAMA 221:1265, 1972
4. Werner SC, Spooner M: A new and simple test for hyperthyroidism employing l-triiodothyronine and the twenty-four-hour I-131 uptake method. Bull NY Acad Med 31:137, 1955
5. Dunn JT, Chapman EM: Rising incidence of hypothyroidism after radioactive-iodine therapy in thyrotoxicosis. N Engl J Med 271:1037, 1964

ADDITIONAL READING

Beierwaltes WH, Wagner HN Jr: Therapy of thyroid diseases with radioiodine. In Wagner HN Jr (ed): Principles of Nuclear Medicine. Philadelphia, Saunders, 1968, pp 343-363

Blahd WH: Scanning of the thyroid gland. In Blahd WH (ed): Nuclear Medicine, 2nd ed. New York, McGraw-Hill, 1971, pp 227-236

Koplowitz JM, Solomon DH: Tests of thyroid function. In Blahd WH (ed): Nuclear Medicine, 2nd ed. New York, McGraw-Hill, 1971, pp 187-226

Means JH, DeGroot LJ, Stanbury JB: The Thyroid and its Diseases, 3rd ed. New York, McGraw-Hill, 1963

Pitt-Rivers R, Trotter R (eds): The Thyroid Gland. London, Butterworth, 1964

Silver S: Radioactive isotopes in clinical medicine. New Engl J Med 272: 466; 515, 1965

Werner SC, Ingbar SH (eds): The Thyroid: A Fundamental and Clinical Text, 3rd ed. New York, Harper and Row, 1971

Workman JB: The thyroid. In Freeman LM, Johnson PM (eds): Clinical Scintillation Scanning. New York, Hoeber Medical Division, Harper and Row, 1969, pp 446-467

chapter 4
KIDNEYS

RADIONUCLIDE RENOGRAM

When first introduced the radionuclide renogram showed great promise of becoming a valuable test of renal function. To the disappointment of those investigators involved with its development and subsequent refinement, it has never received the recognition accorded the other, more frequently employed nuclear medicine studies. This is due in part to the failure to standardize the procedure and to the continued controversy in regard to some aspects of renogram interpretation.

Nevertheless, when properly performed the renogram can be one of the most useful tests of differential renal function. Through external monitoring the activity of a radioactive material is detected and recorded as it travels through the kidney, this record being the renogram tracing.

The radioactive material used is ^{131}I-orthoiodohippurate (Hippuran). It is handled by the kidney in much the same manner as paraaminohippurate (PAH)—principally through tubular excretion and to a lesser extent by glomerular filtration.

An apparatus for performing a renogram is shown in Figure 1 and diagramed in Figure 2. A detector with a 1-inch sodium iodide crystal and flat field collimator is placed over each renal area. These paired detectors are connected to ratemeters and then to rectilinear recorders. A small dose of

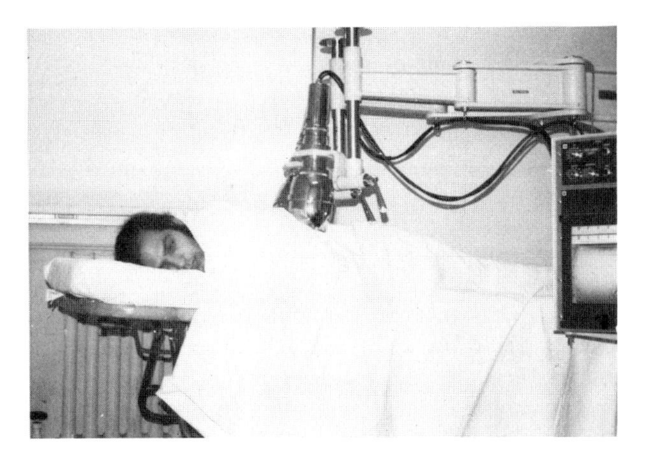

Fig. 1. Apparatus for performing renograms with patient in proper position. The dual probes are directed to the renal areas. Note the elevation of the patient.

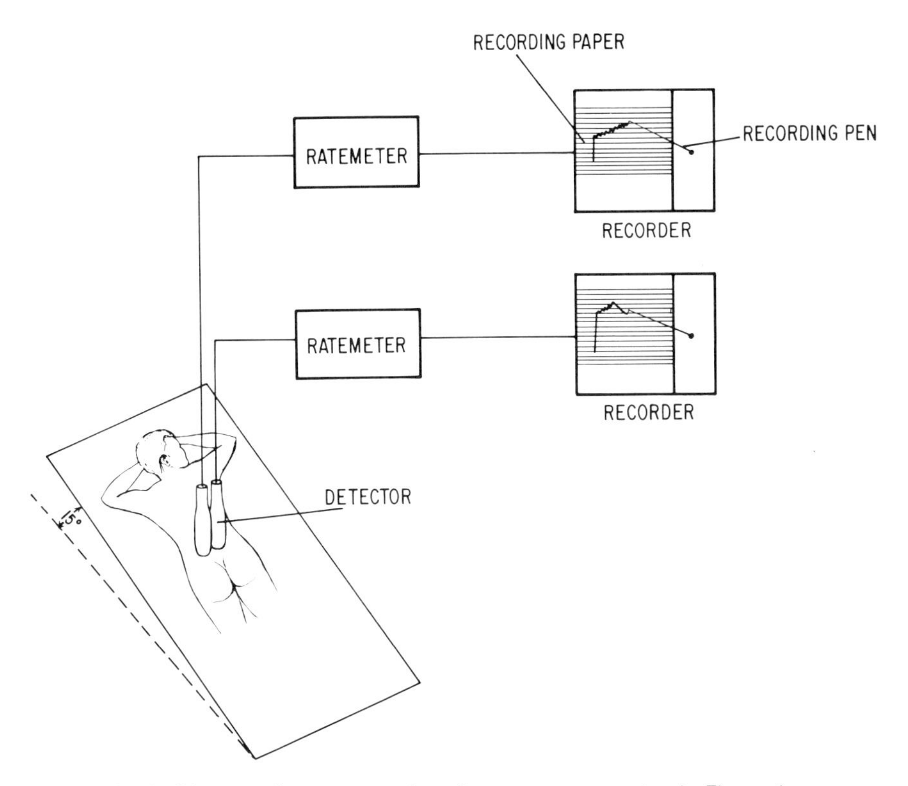

Fig. 2. Diagramatic representation of renogram apparatus in Figure 1.

^{131}I-Hippuran (15 to 25 μCi) is injected intravenously. The radioactivity in the kidneys is detected by each of the paired probes, and the course of the radionuclide in the kidney is reflected on a paper recording or renogram tracing. The paper in the recorder is on a revolving drum, usually moving at a speed of 12 inches/hour.

Preparation for Renogram

It is essential that the renogram be performed under standardized, controlled conditions. If this is not done, the definition of what constitutes a normal renogram becomes meaningless, and subtle abnormalities escape detection. Special attention must be given to hydration and position of the patient and proper placement of the detectors on the renal areas.

Hydration. Extremes of dehydration and overhydration should be avoided. Dehydration results in abnormal prolongation of the transit time and reduced excretion; excess hydration has the opposite effect. In the normal person a urinary output of 1.5 to 7.0 ml/minute is accompanied by reproducible, normal renograms.[1] This may be achieved by having the patient fast for 8 to 12 hours and then drink 750 ml water. The renogram is performed 1 hour later. This is commonly done in the hospital by ordering the patient to be NPO after midnight and scheduling the renogram for between 8 AM and noon.

Position. It is best to perform the renogram with the patient in the recumbent position, either prone or supine. The head of the stretcher should be at a modest elevation (about 15°) so gravity assists urinary drainage. Renograms are often difficult to perform in the sitting position because of patient movement and slumping.

Placing Detecting Probes. The detector must be positioned accurately over each renal area. This is best done by injecting intravenously a very small amount (5 μCi) of ^{197}Hg-chlormerodrin; this enters the renal tubules where it achieves a high concentration. The detectors are then positioned in the areas where the activity from the ^{197}Hg is highest, usually the central portion of each kidney. The ^{197}Hg-chlormerodrin may be injected 30 minutes to 24 hours before localization is attempted. Similar identification of the renal areas can be made from a renal scan, using either radiomercury or ^{99m}Tc-DTPA.

Normal Renogram

The normal renogram has a characteristic pattern consisting of three distinct segments: tracer appearance, buildup phase, and excretory phase (Fig. 3). The ordinate represents radioactivity (counts per minute), and the abscissa time (minutes).

The *tracer appearance* is the initial rise in radioactivity following in-

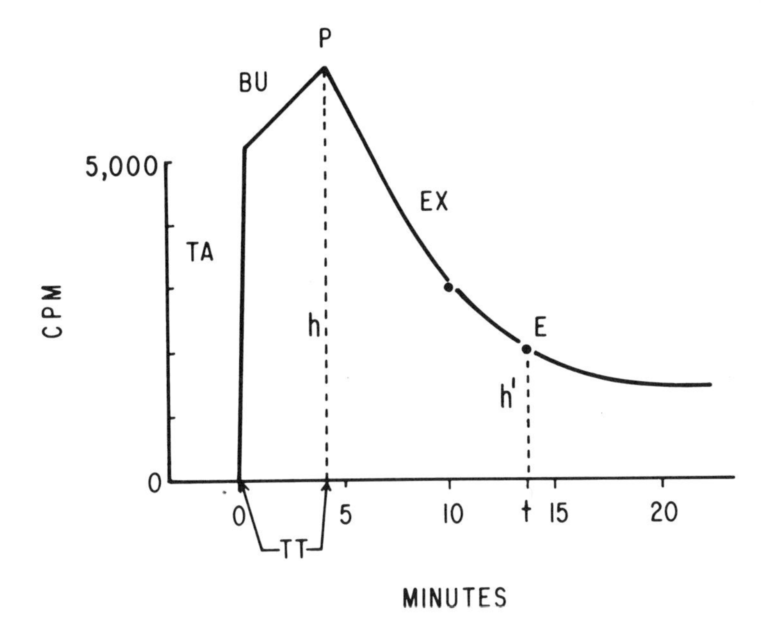

Fig. 3. Normal renogram, CPM, counts per minute, TA, tracer appearance, BU, buildup phase. P, peak. TT, transit time. EX, excretory phase. h, height of peak. E is the point on the excretory curve of height h′ at time t.

travenous injection of the ^{131}I-Hippuran and is due to the presence of the radioactive material in the vasculature within and surrounding the kidney. This phase lasts about 30 seconds. A probe placed over the heart or any major artery would record a similar deflection, but of less amplitude.

Following the tracer appearance there is an upward sloping segment or *buildup phase.* It represents the accumulation of ^{131}I-Hippuran in the renal tubules. In the healthy subject 92 percent of the ^{131}I-Hippuran presented to the tubules is extracted in a single circulation of tracer. The buildup phase continues gradually until the *peak* or point of maximum accumulation is reached.

The *transit time* is the period from the initial upward deflection of the tracer appearance to the point of maximum accumulation or peak. It includes both the tracer appearance and buildup phase, and generally lasts 1.5 to 4.5 minutes.

The *excretory phase* is the downward sloping segment following the peak and represents the discharge of ^{131}I-Hippuran from the renal tubules into the urinary tract. Thus as more radioactive material is lost to the urine, less radioactivity is present in the renal tubules.

The excretion is measured as the percentage drop in radioactivity from the peak at any given time on the excretory curve. If h is the height of the peak and h' the height of a point on the excretory curve at a given time t, then the percentage excretion at time t is represented by the drop from the peak at point E:

$$\% \text{ excretion} = \frac{h-h'}{h} \times 100$$

It is common to measure excretion on the renogram in terms of the time at which the amplitude of the excretory curve is one-half that of the peak, or $T_{1/2}$. This point of 50 percent excretion occurs at approximately 10 minutes after Hippuran injection. However, when comparing the excretory function of the two kidneys it is more meaningful to measure the percent excretion at several intervals. Normally the excretion is at least 40 percent at 7.5 minutes, 50 percent at 10 minutes, and 60 percent at 15 minutes.

Abnormal Renogram

There is no general agreement regarding the characteristics of abnormal renograms. As a matter of convenience, renograms may be classified according to the extent of the renogram abnormality. The following major types of abnormal renograms are listed in order of progressive severity: reduced excretion, prolonged transit time/reduced excretion, "obstructive" pattern, and no-buildup pattern.

Reduced Excretion. An attractive theory holds that in early tubular impairment there is probably normal uptake of ^{131}I-Hippuran but the ability of the renal tubules to discharge it into the tubular urine apparently is reduced. This results in a renogram with a normal transit time but with reduced excretion (Fig. 4). The excretory fall from the peak is less than 50 percent at 10 minutes. This is the mildest form of renogram impairment and is seen, for example, in chronic, inactive pyelonephritis.

Prolonged Transit Time/Reduced Excretion. With continued tubular impairment, the ability to take up as well as to discharge the ^{131}I-Hippuran probably is reduced (Fig. 5). It takes longer for the tracer to enter the tubules, resulting in a prolonged transit time (longer than 4.5 minutes). In addition to the abnormal transit time, the excretion is less than 50 percent at 10 minutes. Such a renogram may be seen in chronic pyelonephritis that is more severe than those cases in which reduced excretion is the only renogram abnormality.

In both the reduced excretion and prolonged transit time/reduced excretion renograms the basic renogram configuration is preserved and the three renogram segments are sharply delineated. As parenchymal disease becomes

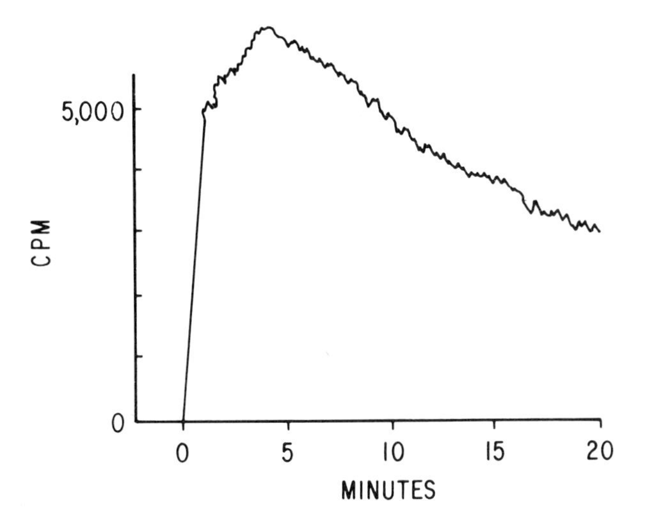

Fig. 4. Reduced excretion renogram. The transit time is normal, but the excretory drop from the peak is only 21 percent at 10 minutes.

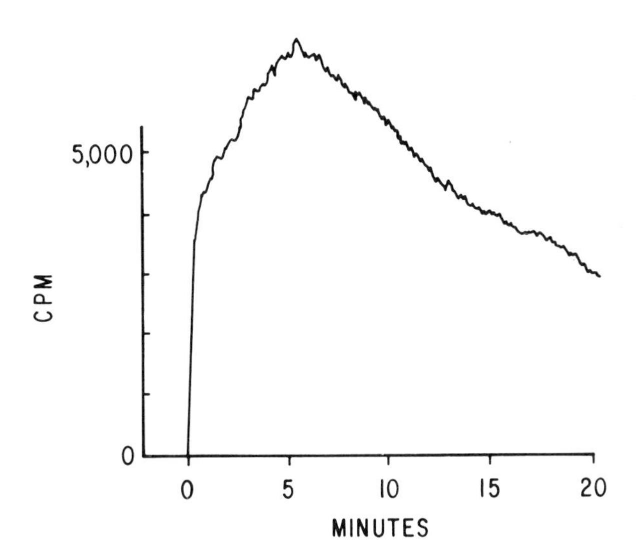

Fig. 5. Prolonged transit time/reduced excretion renogram.

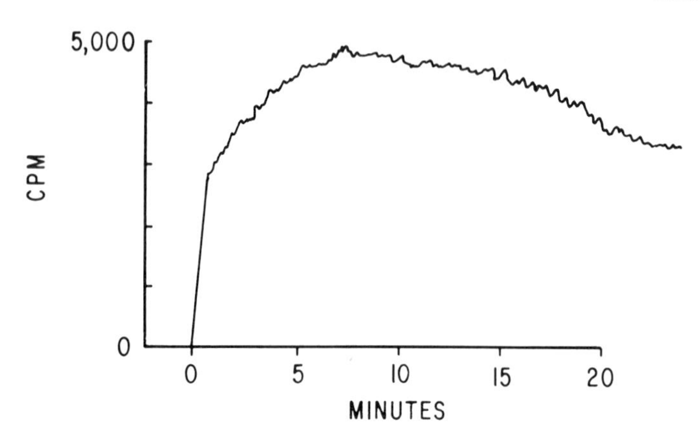

Fig. 6. Prolonged transit time/reduced excretion renogram. Compared with the tracing in Figure 5, this renogram shows more severe excretory impairment and further flattening of the tracing.

more advanced there is even greater prolongation of the transit time and reduction in excretion. The buildup and excretory segments blend, and the tracing appears distorted (Fig. 6). The ability to take up and secrete ^{131}I-Hippuran becomes further reduced, and there is equivalent uptake and discharge of the radioactive material. As a result, a buildup phase is not discernible and the excretory segment is flattened (Fig. 7). Such renograms are

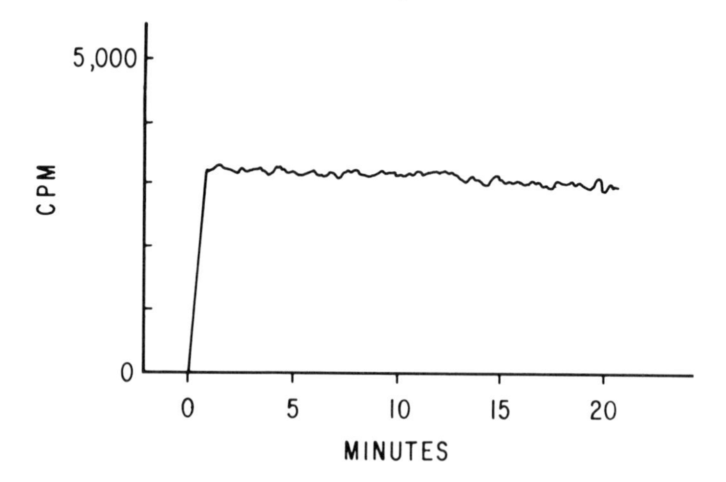

Fig. 7. Loss of basic renogram configuration in patient with chronic, severe parenchymal disease of the kidneys, with flattening of the buildup and excretory phases.

commonly seen in patients with uremia secondary to chronic parenchymal disease; the tracings in these cases are usually similar bilaterally.

"Obstructive" Pattern. Following the tracer appearance, the buildup phase continues to rise and an excretory phase is absent (Fig. 8). The ^{131}I-Hippuran remains and continues to accumulate in the renal tubules because the tubules are unable to excrete it. This may be due to mechanical obstruction to urinary flow (e.g., renal stone, obstruction of a ureter by tumor) or intrarenal obstruction. With the latter, there is blockage of the tubular lumens with sloughed cells, cellular debris, and proteinaceous material. In such cases the ^{131}I-Hippuran continues to accumulate in the already damaged tubular cells, resulting in an "obstructive" pattern. It is especially striking in acute tubular necrosis and acute glomerulonephritis.

As a practical matter, a unilateral obstructive pattern is usually associated with mechanical obstruction. When the obstructive pattern is bilateral, intrarenal obstruction is more likely. The renogram is the most sensitive test to detect urinary obstruction. Obstructive changes are often seen here before they are visualized on the intravenous urogram.

A renogram pattern of intermittent obstruction may be seen when a ball-valve mechanism in the ureter is associated with alternate periods of urinary flow and obstruction. This is best illustrated by a ureteral stone, which results

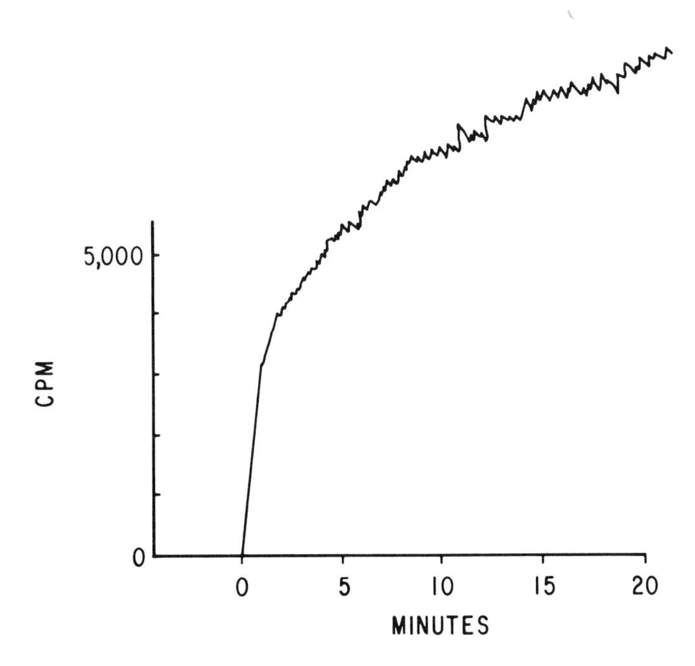

Fig. 8. "Obstructive" pattern renogram.

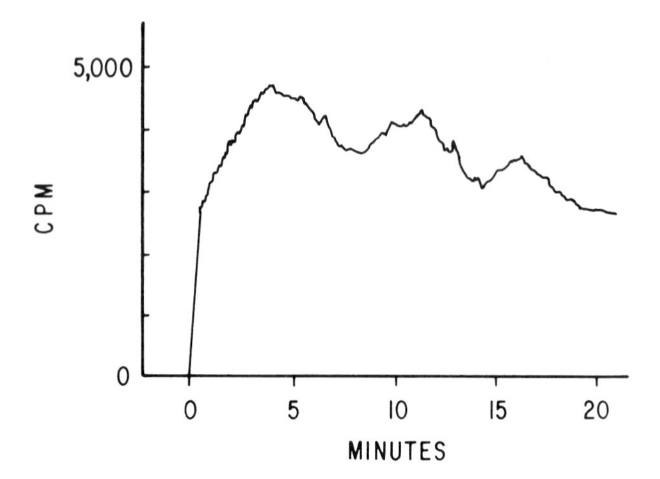

Fig. 9. "Intermittent obstruction" pattern renogram. This may be associated with a ball-valve mechanism, such as a ureteral stone.

in urinary obstruction at the time of ureteral peristalsis but does not interrupt urinary flow during periods of ureteral relaxation (Fig. 9). Ureteral spasms, such as those associated with ureteral catheterization, result in similar renogram abnormalities.

No-Buildup Pattern. There is absence of a second segment or buildup phase (Fig. 10) in the pattern seen with nephrectomy or nonfunction. Following the tracer appearance there is a downward sloping segment that

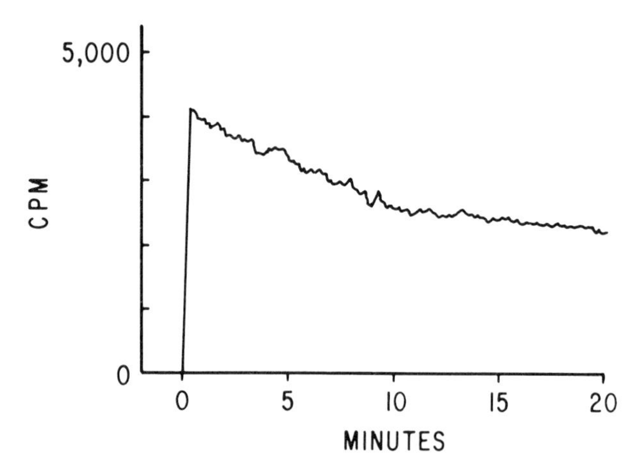

Fig. 10. No-buildup pattern renogram.

represents the gradual reduction in radioactivity in the blood as the ^{131}I-Hippuran is taken up in the contralateral kidney. The downward slope in this type of renogram is *not due to excretion.* The tracing is similar to but with less amplitude than that which would be seen if the detector was placed over the heart.

A no-buildup pattern may also be seen in those cases in which there is some renal function left but the tubular damage is so extensive that little or no tracer can be taken up. Still another instance in which a no-buildup pattern may be seen is when the detector is improperly placed because of faulty kidney localization and does not face renal tissue. This is most commonly encountered with ptosis of a kidney.

Unilateral Renal Artery Stenosis and Hypertension

One of the most frequent applications of the renogram has been its use as a screening test to detect those patients with hypertension due to unilateral renal artery stenosis and ischemia. This surgically correctable condition is thought to be present in as many as 5 percent of all patients with high blood pressure.

The renogram changes that occur in unilateral renal ischemia are more easily understood if the cardinal functional and histologic alterations that occur in the kidney in this disease are kept in mind. After narrowing of a renal artery, usually by an atheromatous plaque, there is a reduction in blood flow to the affected kidney. The ischemic kidney elaborates renin, resulting in hypertension. If the stenosis is relieved or the kidney removed in time, the elevated blood pressure may return to normal.

The pathologic lesion in the kidney with reduced blood flow is ischemic atrophy of the renal tubules. The extent of the tubular damage is related to the duration and degree of renal artery constriction and reduction in renal blood flow.

Functional changes also occur in the ischemic kidney. There is increased tubular reabsorption of water and sodium in the kidney with reduced blood flow compared to the contralateral uninvolved kidney. This results in reduced urinary excretion on the side of the affected kidney.

The renogram findings associated with unilateral renal ischemia are probably related to both the anatomic changes of tubular atrophy and the functional changes of increased water reabsorption and reduced urinary excretion. The extent of these changes is reflected in the degree of severity of renogram abnormalities. The theoretical considerations centering around the renogram in renovascular hypertension are indeed many, [2-4] the following of which is only one.

During the early stages of unilateral renal artery stenosis in which the reduction in renal blood flow is not marked, the pathologic changes of ischemic atrophy are not yet manifest. However, there is reduced urinary output on the affected side, which is reflected on the renogram by reduced excretion in an otherwise normal tracing. As the renal artery stenosis and ischemia progress, the tubules begin to undergo pathologic changes of ischemic atrophy and there is continued reduction in urinary excretion. The reduced ability of the tubules to take up the ^{131}I-Hippuran coupled with the decreased urinary output results in a renogram with both a prolonged transit time and reduced excretion (Fig. 11).

More advanced ischemic tubular atrophy results in sloughing of tubular cells and the presence of cellular debris and proteinaceous material in the tubular lumens. This plus the marked reduction in urinary excretion in the ischemic kidney can result in an "obstructive" pattern of the intrarenal variety. Finally, unilateral renal artery stenosis and ischemia may be severe and of long duration. In these cases there is such a miniscule quantity of functioning tubules that the renogram shows a no-buildup pattern.

From this great variety of renogram abnormalities, it is evident there is no characteristic renogram abnormality in unilateral renal artery stenosis and ischemia. Any type of renogram abnormality may be encountered, depending on the severity and duration of the ischemia.

How, then, does one use the renogram as a screening procedure to detect those patients with surgically correctable renovascular hypertension? If the renogram is normal it practically rules out the possibility of unilateral renal ischemia as a cause of the hypertension, and further work-up with renal angiography is not warranted in the absence of other convincing evidence of unilateral renal artery stenosis.

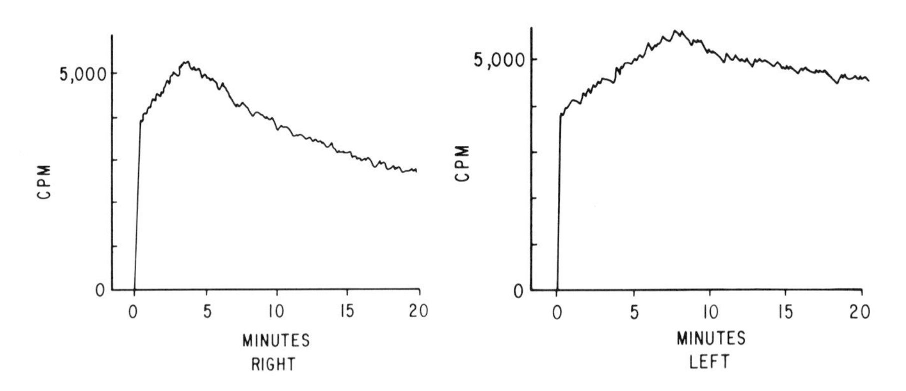

Fig. 11. Renogram of patient with hypertension associated with left renal artery stenosis showing prolonged transit time and reduced excretion on the affected side. Note the disparity between the two tracings, the right renogram having a normal transit time but reduced excretion.

The possibility of unilateral renal artery stenosis and ischemia is considered in the patient whose renogram demonstrates two features: an abnormality and a disparity between the left and right tracings. If these two factors are manifested on the renogram, further work-up with selective renal angiography should be performed. It is from these cases that the candidate with surgically correctable hypertension may be selected.

As stated earlier, there are countless other renal diseases that result in abnormal renograms. If there is no disparity between two abnormal renograms in the hypertensive patient, it is more likely that the abnormalities are associated with bilateral renal disease.

The renogram thus becomes one of the many factors in evaluating the patient with possible unilateral renal ischemia and falls in place along with plasma renin determinations, the intravenous urogram, renal angiography, and the specific studies of differential renal function (Howard and Stamey tests).

Other Clinical Considerations

The renogram may be used to assess differential and individual renal function in any situation in which this information is desired. The renogram, however, does not provide both anatomic and functional data as does the intravenous urogram; it is limited to the area of renal function. However, the renogram does reflect the functional status of each kidney in a somewhat quantitative manner.

Its use as a test of differential renal function is frequently applied in the evaluation and follow-up of patients with suspected or known renal disease. Once initial tracings have been obtained, the progress of a patient with a renal problem may be followed by performing serial renograms, usually at intervals of 3 to 6 months. The renogram of a patient with chronic pyelonephritis is shown in Figure 12. Both kidneys here show impairment of renal function, but it is more severe on the right. Such semiquantitative information is difficult to obtain from the intravenous urogram. In addition, the radiation exposure from the ^{131}I-Hippuran renogram is 1 percent of that obtained from intravenous urography.[5]

Because the renogram is the most sensitive test to demonstrate obstruction in the urinary tract, it is often used to follow patients treated for pelvic and abdominal neoplasms—to detect obstruction from tumor. Since the procedure is so simple, it is easier to have the patient undergo renography at intervals of 3 to 4 months rather than have roentgenologic studies. In a patient with an unexplained elevation of blood urea nitrogen, the renogram may be used to explore the possibility of a urinary obstruction as a cause of the azotemia.

The renogram may be used to demonstrate differential renal function in

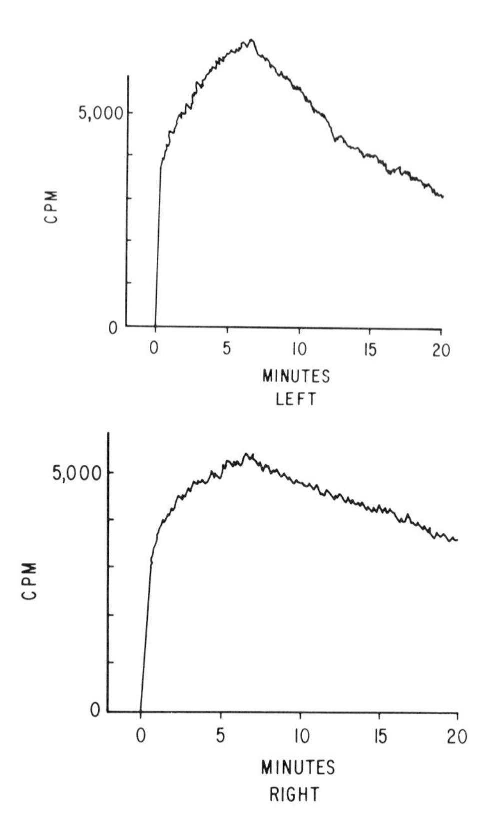

Fig. 12. Renogram of patient with chronic pyelonephritis. The transit time is prolonged in both tracings (7 minutes on each side). The excretory drop from the peak at 10 minutes is 13 percent on the right side and 20 percent on the left, clearly indicating better excretory function in the left kidney.

those patients in whom intravenous urography is contraindicated because of sensitivity to the contrast material. Renography is often the principal method by which these patients are studied for evidence of renal impairment. The renogram of a patient with such an allergy and carcinoma of the bladder is illustrated in Figure 13. He developed azotemia, and the renogram showed an "obstructive" pattern on the left side. The obstruction was found at the left ureterovesical orifice, and the left ureter was reimplanted in the bladder. At 3 months after surgery the renogram shows a return to normal configuration on the left side, although there is some reduction in excretion.

The renogram may be performed at the bedside. In those patients too ill to be moved or prepared for intravenous urography, the renogram may provide sufficient information regarding the functional status of the kidneys.

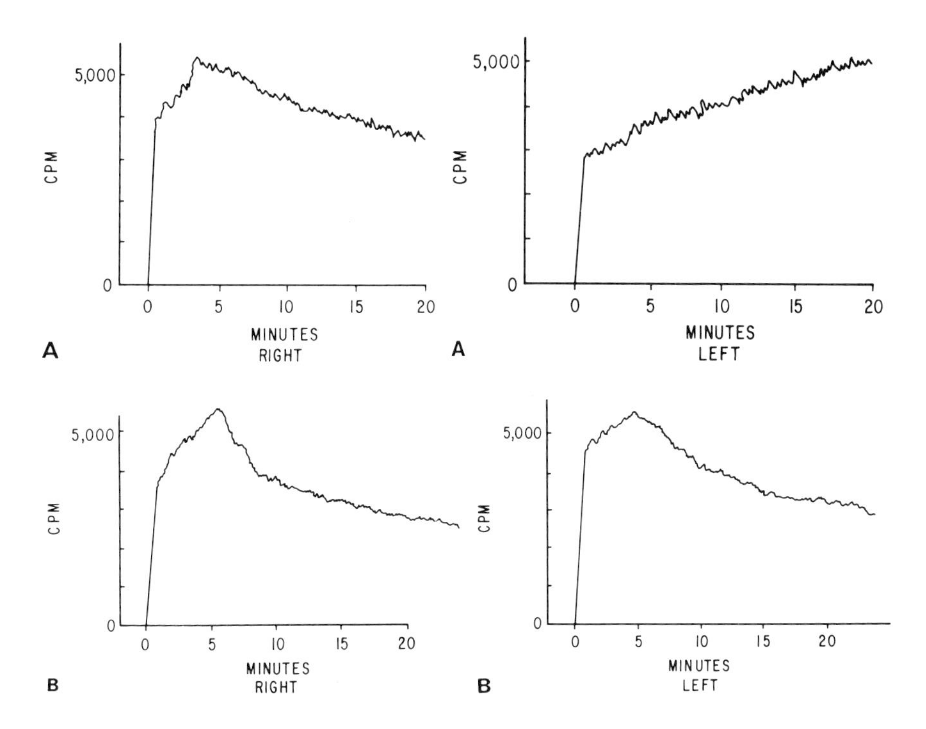

Fig. 13. Renogram showing "obstructive" pattern on the left side because of tumor blocking the left ureterovesical orifice. B. Renogram of same patient 3 months after the left ureter was reimplanted into the bladder. The "obstructive" pattern is no longer present and the three renogram segments are now clearly delineated.

Of course in special situations in which controlled hydration and urinary output is impossible or undesirable, useful information of differential renal function can still be supplied by the renogram. This is especially true in regard to the presence or absence of a nonfunctioning kidney or urinary obstruction.

RENAL IMAGING

Renal imaging with radionuclides is rarely able to provide the degree of information obtained with roentgenologic studies of the kidney, such as the intravenous urogram and the renal angiogram. Nevertheless, there are many situations in which imaging with radioactive materials is helpful and desirable, such as demonstrating renal size and position and identifying the kidney's borders prior to percutaneous biopsy.

A wide variety of radiopharmaceutical agents have been used to visualize the kidney with tracers. At the present time, renal imaging may be performed with one of three types of agents.

Radioactive Chlormerodrin

Mercurial diuretics enter the renal tubule cells where a high concentration of mercury can be achieved. The radiopharmaceutical agent may remain in these cells for several hours prior to excretion.[6] ^{203}Hg-chlormerodrin was the first radioactive material to be widely used as a renal scanning agent. Because of the long half-life (46 days) and relatively high gamma emission (279 keV) it has been largely supplanted by ^{197}Hg-chlormerodrin. The latter has a half-life of 65 hours and a principal gamma emission of 77 keV.

The usual adult dose of ^{203}Hg-chlormerodrin is 100 μCi, and for ^{197}Hg-chlormerodrin it is 200 μCi. In the healthy subject there is sufficient concentration of the radioactive material in the renal tubules as early as 2 hours after intravenous administration to permit imaging at that time. However, some prefer to wait until 3 to 4 hours have elapsed before imaging is attempted.

Tagged Chelates

During the past 5 years, a number of chelated compounds have become available for radionuclide investigation of the kidneys. Since true chelates are not metabolized and are excreted exclusively by glomerular filtration, the tagging of these substances with tracers has found ready application in renal studies.

The chelating agent that has been used most widely is diethylenetriaminepentaacetic acid (DTPA) and ^{99m}Tc is most often the tracer. Since ^{99m}Tc must be in the tetravalent state to form a stable chelate with DTPA, a reducing agent must be used for the conversion of ^{99m}Tc at the VII oxidation state (as pertechnetate) to the oxidation state of IV.[7] Among the reducing agents that have been employed are the stannous ion and ferric ascorbate.

When the reduction process is carried out with stannous ion, the resulting ^{99m}Tc-DTPA (Sn) is an extremely stable compound and has been employed for renal imaging following the intravenous administration of 5 to 10 mCi. However, since this material is excreted so rapidly by the kidneys through glomerular filtration, to accomplish imaging of the renal cortex it is usually necessary to perform the procedure during the first 15 minutes following tracer injection.

The complex obtained when ferric ascorbate is the reducing agent is ^{99m}Tc-iron-ascorbate-DTPA. This material is not stable in vivo, is not a true chelate, and is not excreted exclusively by glomerular filtration. Its mechanism of localization is probably similar to that of a related compound, ^{99m}Tc-ferrous ascorbate. With the latter, it has been postulated that a portion of the ^{99m}Tc becomes bound to protein in renal tubular cells.[8] This is because the stability of the ^{99m}Tc-sulfhydryl complex is much greater than that of the ^{99m}Tc-ferrous ascorbate.[8]

Similar protein binding by ^{99m}Tc (probably from 10 to 20 percent) most likely occurs with ^{99m}Tc-iron-ascorbate-DTPA. Because of temporary retention of ^{99m}Tc in the cells of the renal tubules, imaging is usually performed 1 to 2 hours following the injection of 3 to 5 mCi. The technetium 99m images of kidneys illustrated in this chapter were all performed with ^{99m}Tc-iron-ascorbate-DTPA.

^{131}I-Orthoiodohippurate (Hippuran)

Hippuran is actively secreted by the renal tubules, and its use in renal imaging is usually reserved for those instances in which renal impairment is so severe that sufficient radioactivity cannot be attained with the other renal scanning agents.[9] The usual adult dose is 100 to 200 μCi. When ^{131}I-Hippuran is employed, frequent imaging is attempted during the 30 minutes following its intravenous injection so the renal tubular areas can be visualized before the majority of radionuclide is excreted and present in the renal pelvis.

Normal Renal Scan

Renal imaging is usually performed in the posterior position (Fig. 14). It is common to see increased radioactivity of very low intensity superior to the right kidney; this simply represents circulating radionuclide in the liver. The size and position of the kidney can be readily determined from the study.

Abnormal Renal Scan

Space-occupying Lesions. Localized destruction and replacement of nephrons are reflected on the scan as areas of absent radioactivity. Thus cyst, tumor, or intrarenal abscess may have similar appearances. Although the location and configuration of these regions of absent radioactivity may in

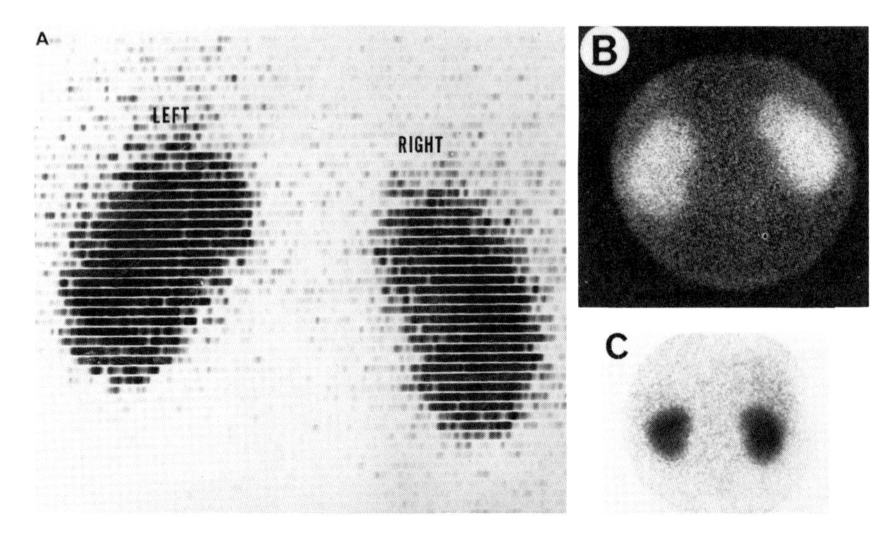

Fig. 14. Normal renal imaging performed in the posterior position. A. Rectilinear scan following ^{197}Hg-chlormerodrin administration. B and C. Scintillation camera images after ^{99m}Tc-DTPA injection using a parallel hole collimator (B) and a diverging collimator (C).

themselves offer diagnostic clues, it is usually unrewarding to attempt to identify precisely the nature of the space-occupying lesion from the renal scan abnormality alone.

In some cases the space-occupying lesion may be well delineated on the scan (Figs. 15 and 16). In others there may be extensive destruction of renal tissue so that imaging of the involved kidney suggests the presence of an organ of reduced size.

Position. Scanning is often helpful in the demonstration of ptosis, size (Fig. 17), and renal displacement because of an extrinsic mass. Not in-

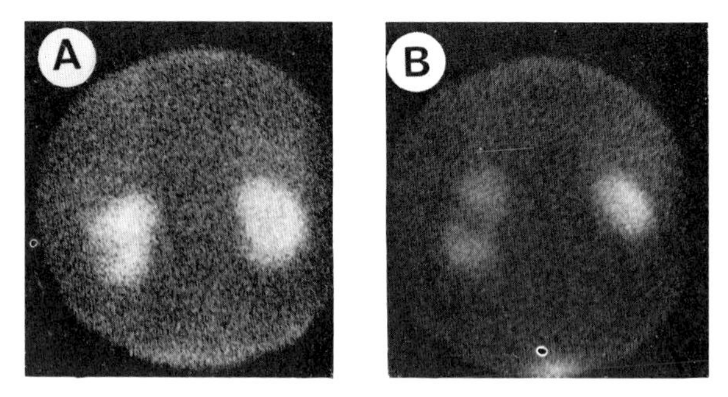

Fig. 15. Absent radioactivity in the left kidney associated with a renal cyst (^{99m}Tc-DTPA). A. Rounded area, left lower pole. B. Mid-portion, left kidney.

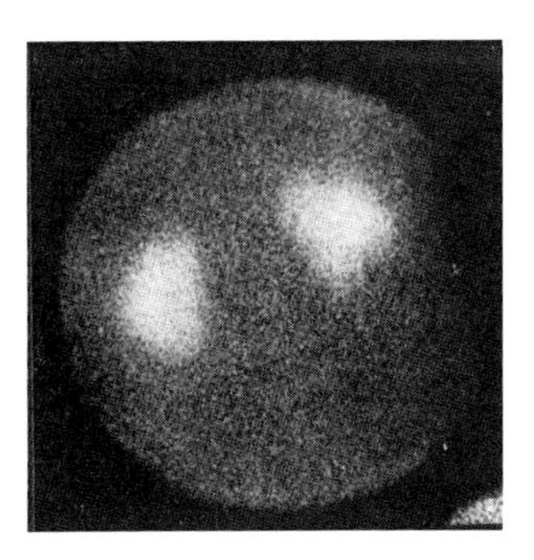

Fig. 16. Absent radioactivity in the lower portion of the right kidney due to tissue destruction associated with a hypernephroma (^{99m}Tc-DTPA).

frequently only one kidney is identified on a roentgenologic study, and the renal scan is employed to determine the presence or absence of the other kidney (Fig. 18). If viable renal tissue is present the structure probably can be demonstrated with radionuclide imaging. Scanning is often performed to identify the kidneys before percutaneous renal biopsy.

Congenital Abnormalities. The renal scan has been employed in many cases to assist in evaluating congenital abnormalities. Thus radionuclide im-

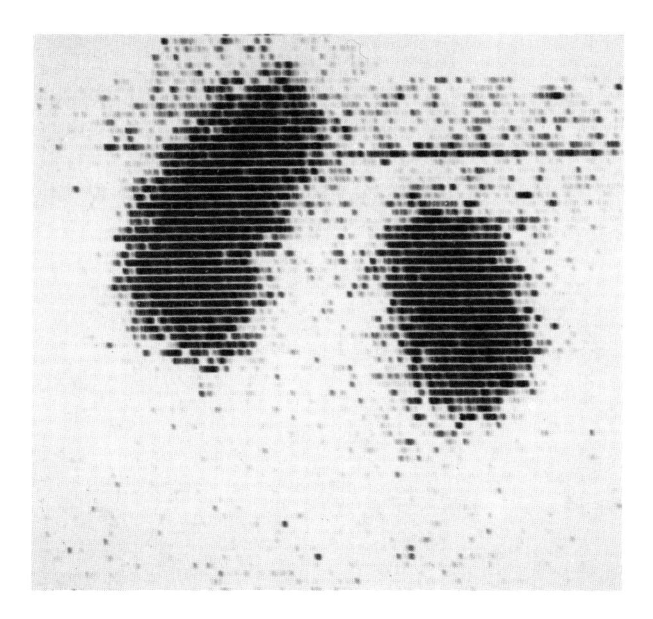

Fig. 17. Marked reduction in size of right kidney demonstrated on rectilinear scan using ^{197}Hg-chlormerodrin.

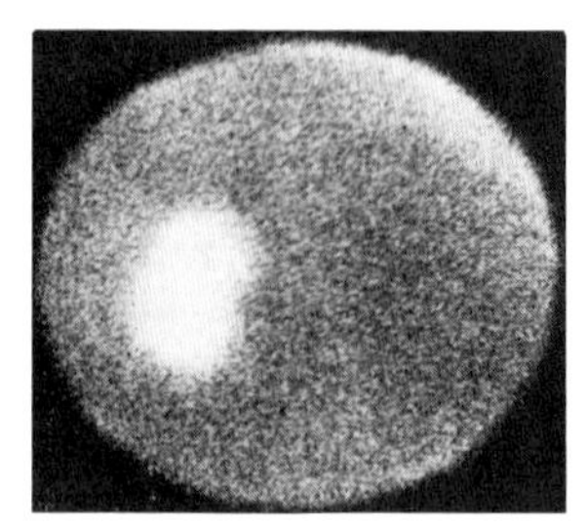

Fig. 18. Failure to visualize the nonfunctioning right kidney after ^{99m}Tc-DTPA injection.

aging is helpful in determining whether the site of fusion is superior or inferior in patients with horseshoe kidneys (Fig. 19).

Clinical Considerations

In patients sensitive to contrast materials, renal imaging with radioactive materials may provide the only means by which the kidneys can be visualized. This is probably one of the most important applications of the renal scan. However, as indicated earlier it is uncommon for renal scanning to offer any anatomic information beyond that obtained through urography or angiography.

If renal function is impaired to the degree that little or no radioactive material is made available to the kidney for tubular uptake or glomerular filtration, images of the liver and spleen (rather than the kidneys) are obtained as the circulating level of radionuclide continues to be high. This is especially true when radioactive chlormerodrin is used. If there is only modest elevation of the blood urea nitrogen, sufficient ^{197}Hg- or ^{203}Hg-chlormerodrin might accumulate in the tubules 24 hours after tracer administration so that

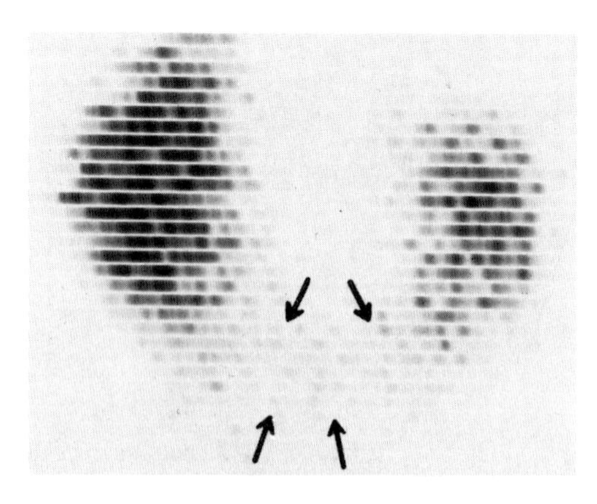

Fig. 19 Rectilinear scan demonstrating site of fusion (arrows) at the lower poles (^{197}Hg-chlormerodrin).

imaging can be accomplished. In such cases, imaging with ^{131}I-Hippuran can be helpful. A prior renogram determines the time at which there is maximum tubular uptake of radionuclide, this being the ideal time for imaging.

REFERENCES

1. Wedeen RP, Goldstein MH, Levitt MF: The radioisotope renogram in normal subjects. Am J Med 34:765, 1963
2. Taplin GV, Dore EK, Johnson DE: The quantitative radiorenogram for total and differential renal blood flow measurements. J Nucl Med 4:404, 1963
3. Luke RG, Briggs JD, Kennedy AC, et al: The isotope renogram in the detection and assessment of renal artery stenosis. Q J Med 35:237, 1966
4. Maxwell MH, Lupu AN, Taplin GV: Radioisotope renogram in renal arterial hypertension. J Urol 100:376, 1968
5. Henk JM, Cottrall MF, Taylor DM: Radiation dosimetry of the ^{131}I-Hippuran renogram. Br J Radiol 40:327, 1967
6. McAfee JG, Wagner HN Jr: Visualization of renal parenchyma by scintiscanning with HG^{203}Neohydrin. Radiology 75:820, 1961
7. Eckleman WC, Meinken G, Richards P: The chemical state of ^{99m}Tc in biomedical products. II. The chelation of reduced technetium with DTPA. J Nucl Med 13:577-581, 1972
8. Reba RC, Poulose KP, Kirchner PT: Radiolabeled chelates for visualization of renal function and structure with emphasis on their use in renal insufficiency. Semin Nucl Med 4:151-168, 1974
9. Freeman LM, Goldman SM, Shaw RK, et al: Kidney visualization with ^{131}I-o-iodohippurate in patients with renal insufficiency. J Nucl Med 10:545, 1969

ADDITIONAL READING

Blaufox MD (ed): Progress in Nuclear Medicine: Evaluation of Renal Function and Disease with Radionuclides. Baltimore, University Park Press, 1972

Blaufox MD, Funck-Brentano J-L (eds): Radionuclides in Nephrology. New York, Grune and Stratton, 1973

Britton KE, Brown NJG: Clinical Renography. Chicago, Year Book, 1973

Farmelant MH, Burrows BA: The renogram: physiologic basis and current clinical use. Seminars Nucl Med 4: 61-73, 1974

Kolar M: Radioisotope Renography. Acta Univ Carol Med Praha Monograph 48, 1971

Taplin GV, Nordyke RA: Radioisotope renography. In Blahd WH (ed): Nuclear Medicine, 2nd ed. New York, McGraw-Hill, 1971, pp 402-415

Timmermans L, Merchie G (eds): Radioisotopes in the Diagnosis of Diseases of the Kidneys and Urinary Tract. Amsterdam, Excerpta Medica, 1969

Wagner HN Jr, Reba RC, Goodwin DA: The kidney. In Wagner HN Jr (ed): Principles of Nuclear Medicine. Philadelphia, Saunders, 1968, pp 628-654

Tauxe WN: Renal scanning. In Blahd WH (ed): Nuclear Medicine, 2nd ed. New York, McGraw-Hill, 1971, pp 402-415

chapter 5

LUNGS

Radionuclide studies of the lung for the most part have centered on pulmonary perfusion scans, used for the possible detection of areas of pulmonary embolization. In addition to visualizing the pulmonary arterial distribution, radioactive materials have been useful in assessing and imaging the functioning portions of the respiratory tree and in performing pulmonary function studies.

PULMONARY PERFUSION STUDIES

Studies of the pulmonary arterial circulation (perfusion studies) are performed following intravenous injection of one of the following types of radioactive materials: (1) a tracer tagged to a particle whose diameter is larger than that of a pulmonary capillary, such as ^{131}I-macroaggregated serum albumin and ^{99m}Tc-albumin microspheres; or (2) a radioactive gas that is insoluble in blood and is ultimately expired in the lungs, such as xenon 133.

Imaging With Large Particles

The average diameter of a human pulmonary capillary is approximately 7 μ. If a tagged material with a diameter larger than that is injected into a peripheral vein, it travels to the right heart and pulmonary artery and finally becomes lodged in the pulmonary capillaries and precapillary arterioles.[1]

A number of large particles tagged with a tracer have been employed for lung imaging. Among these are ^{131}I-macroaggregated serum albumin (MAA), ^{99m}Tc-albumin microspheres, and iron hydroxide aggregates labeled with ^{99m}Tc or ^{113m}In If there has been no interruption or blockade in the pulmonary arterial system, intravenous administration of each of these radionuclides results in fairly uniform distribution of radioactive material, as these large particles are filtered out in the pulmonary capillaries and precapillary arterioles. On the other hand, if a pulmonary artery or one of its branches is occluded, as with a pulmonary embolus, the tagged material is unable to enter the capillaries supplied by the involved artery and radioactivity is then absent from this region. Thus the pulmonary perfusion scan shows an area of absent radioactivity corresponding to the pulmonary capillary distribution of the occluded vessel.

The radioactive particles that become lodged in the pulmonary capillaries are themselves microemboli. When performing a perfusion scan with large particles, about 0.22 percent of the pulmonary capillaries become occluded.[2] However, this introduces no hazard or untoward symptoms for the patient.

Radionuclides. Several radioactive materials are currently in use for pulmonary perfusion scanning.

^{131}I-MAA. This was the first material to be successfully employed in lung scanning on a wide basis. Prepared through heating and pH adjustment, these aggregated protein particles range in size for the most part from about 25 to 75 μ. The radionuclide undergoes degradation approximately 4 to 6 hours after injection as the ^{131}I is separated and the protein undergoes lysis into smaller units. The ^{131}I is taken up in part by the thyroid, despite prior administration of Lugol's solution, and the remaining portion is excreted, primarily through the kidneys. The fate of the protein is still not fully understood.[3] There is evidence that after the protein is broken down into smaller units, the particles are phagocytized primarily in the Kupffer cells in the liver but also elsewhere in the reticuloendothelial system (RES), in the spleen and bone marrow. The usual adult dose is 300 μCi.

^{99m}Tc-Albumin Particles. This is probably the most commonly used radionuclide for perfusion imaging of the lungs today. The albumin microspheres or MAA are of uniform size and are about 30 μ in diameter; they are tagged with ^{99m}Tc, the usual adult dose being approximately 3 mCi. The albumin particles apparently break down into smaller units and are handled in a manner similar to that described for MAA.

Tagged Iron Hydroxide Aggregates. Iron hydroxide aggregates, ranging in particle size from about 10 to 30 μ, may be tagged with ^{99m}Tc or ^{113m}In The fate of the material is still not entirely clear and apparently varies with the label employed.[3] There is evidence that a portion of the iron hydroxide remains in the lungs for several days, while in the preparations labeled with ^{113m}In varying amounts of the tracer are phagocytized in the RES. Some of the iron is incorporated into erythrocytes. The usual adult dose for the ^{99m}Tc-iron hydroxide aggregates is about 1 mCi and for the ^{113m}In preparation 3 mCi.

NORMAL LUNG SCAN

The radionuclide is injected intravenously with the patient in the supine position so gravity may assist the tagged particles in traveling to the relatively poorly perfused apical regions of the lungs. Imaging may be commenced almost immediately after injection.

Imaging is generally performed in the anterior, posterior, and both lateral positions. A normal set of images is shown in Figure 1. There is un-

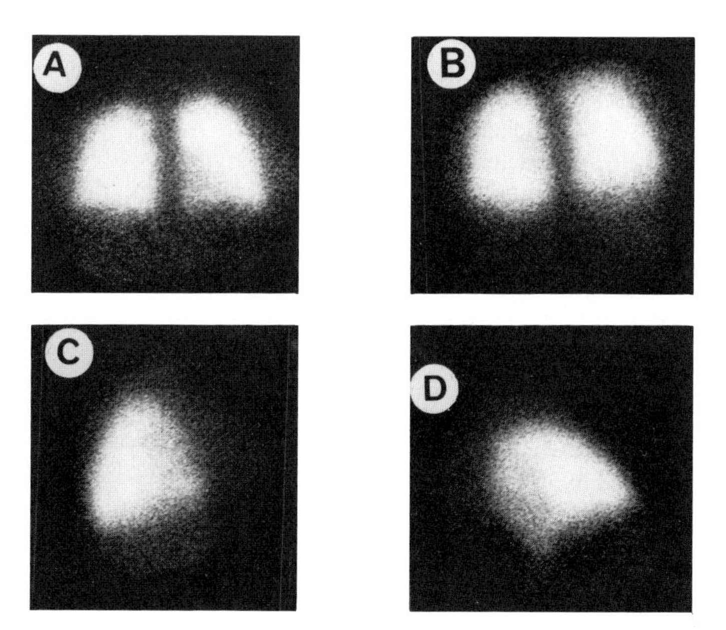

Fig. 1. Normal pulmonary perfusion images in the anterior (A), posterior (B), right lateral (C), and left lateral (D) positions. The study was performed on a scintillation camera following administration of 3 mCi ^{99m}Tc-albumin microspheres. Note the cardiac impression in the anterior position.

iform pulmonary arterial perfusion throughout, except for the minor apical reduction. However, because of the larger mass centrally in each lung, the radioactivity here appears greater than at the periphery in the anterior and posterior positions. The cardiac area is well outlined, especially in the anterior and left lateral positions.

ABNORMAL LUNG SCAN

Pulmonary Embolization

The most frequent application of pulmonary perfusion scanning has been to detect areas of pulmonary embolization. Once an embolus is lodged in a pulmonary artery or one of its branches, there is a reduction or absence of perfusion distal to the clot. If the tracer is injected intravenously it is blocked from entering the involved vessel, and this is manifested on the lung scan as an area of absent radioactivity.

Probably the occlusion need not be complete for the embolization to show up on the lung scan. If there is an incomplete occlusion, some pulmonary arterial perfusion still occurs, but little or no radioactive material enters this vessel because of the reduced blood flow to it; as a result, there is an area of absent radioactivity seen on the scan.

There is no characteristic perfusion scan appearance for a region of pulmonary embolization. The area in which the absent radioactivity is detected is dependent on the size of the occluded vessel and the site at which the embolus is lodged. Thus an entire lobe may be involved or only a small peripheral segment. Central areas *alone* of absent radioactivity are not commonly seen with pulmonary embolization.

The lung scan of a patient with pulmonary embolization involving the right upper and lower lung fields is seen in Figure 2A. The scan was performed 2 days after an episode of left anterior chest pain; at the time of the study the patient was no longer symptomatic and the chest x-ray showed no evidence of lung disease (Fig. 2B). The patient continued to do well, and the scan 2 weeks later showed a return toward normal with reperfusion of the previously involved areas of embolization (Fig. 2C). This case illustrates several important characteristics in the perfusion scan appearance of pulmonary embolization. Among them are the patient's initial left-sided symptoms but subsequent right-sided scan findings, and evidence of reperfusion only 1 week after demonstration of the area of embolization.

Clinical Suspicion, Normal Chest X-Ray, Area of Absent Perfusion. If there is clinical suspicion of a pulmonary embolus and the chest x-ray shows no lung abnormality, the presence of an area of absent

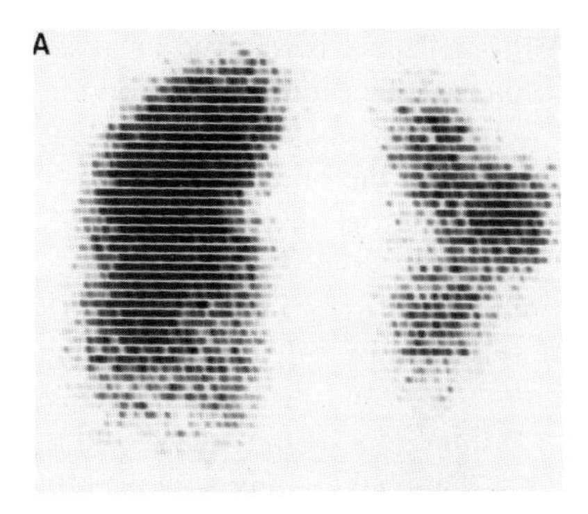

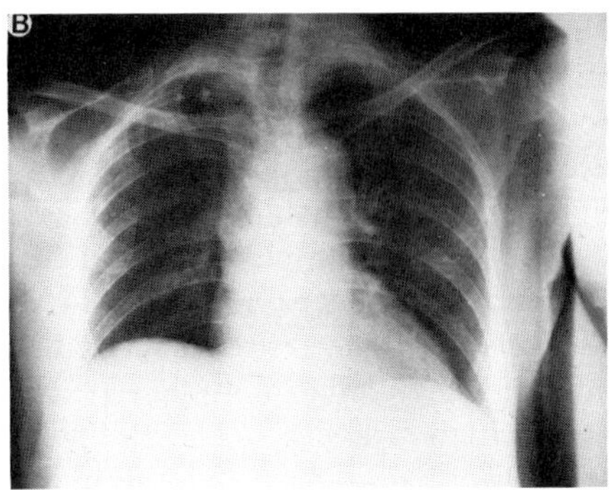

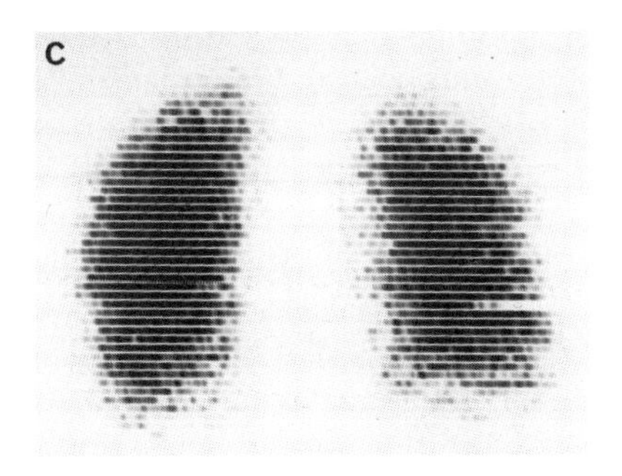

Fig. 2. A. Posterior lung scan showing absent pulmonary arterial perfusion in the right upper and lower lung fields, findings consistent with pulmonary embolism. B. The chest x-ray is normal. C. A repeat scan 2 weeks later shows reperfusion in the involved areas. ^{131}I-MAA was used for scanning.

pulmonary arterial perfusion on the scan is highly suggestive of pulmonary embolization. This is the situation in which the lung scan has the most credence when the diagnosis of embolization is entertained. In fact, many authorities hold that the lung scan findings in such a situation can establish a diagnosis of pulmonary embolization. It is interesting that the area of involvement is often at a site different from where the patient's symptoms are located.

Reperfusion. It is frequently (but not always) characteristic for areas of pulmonary embolization to exhibit reperfusion a short time after arterial occlusion has occurred (5 to 7 days). It is probable that as the clot undergoes organization or fragmentation the blood flow to the involved vessel is reestablished; this is manifested on the scan as a return to normal or improved radioactivity.[4] Areas of absent perfusion associated with underlying pulmonary disease usually have constant perfusion defects, which demonstrate little or no change on serial scans.

Involvement at Other Sites. It is not uncommon to find both reperfusion in the area of the initially imaged abnormality and a region (or regions) of absent perfusion, suggesting pulmonary embolization elsewhere. This often occurs in the absence of recurrent symptoms. These areas of absent radioactivity usually show the same phenomena of reperfusion noted earlier.

Perfusion and Ventilation Studies. Although there is reduced or absent pulmonary arterial perfusion in the region of embolization, ventilation may be normal. This may be demonstrated on ventilation scans using a radioactive gas (see Ventilation Studies in this chapter).[5] On the other hand, a perfusion defect accompanied by a corresponding ventilatory impairment usually signifies underlying pulmonary disease other than embolization.

Pulmonary Infarction. If a clot continues to obstruct a vessel in the pulmonary arterial tree, it is likely that infarction will result. In such a situation, rather than exhibiting reperfusion on the scan the area of absent radioactivity persists. These findings continue to some degree until recanalization has occurred; this may take place approximately 8 weeks after lodging of the embolus, although sometimes it takes 3 to 4 months. However, infarction is an extremely rare sequela of pulmonary embolization.

Chest X-Ray. Never attempt to interpret a lung scan in the absence of a chest x-ray. Countless entities may be identified on chest x-rays whose scan appearance may mimic the findings seen with pulmonary embolization.

Large areas of absent pulmonary perfusion are sometimes seen on scans in situations in which there are multiple pulmonary emboli. An example is the case of a 56-year-old man hospitalized because of severe left-sided chest pain that was at first thought to be associated with a myocardial infarction. The patient was asymptomatic 3 days after the initial episode, and at the time of the initial scan (Fig. 3A) only a portion of the right lower lung showed evidence of pulmonary arterial perfusion. The chest x-ray did not show such massive perfusion defects (Fig. 3B).

It is unlikely that these findings infer an absence of perfusion to the affected areas because such a situation would probably be incompatible with life. There has, however, been a major reduction in pulmonary arterial flow to the involved regions, as evidenced by angiographic findings of pulmonary emboli at the origins of the pulmonary arteries supplying these areas, and the radioactive material that entered this arterial distribution is insufficient to be detected. Nevertheless, the areas of embolization are indeed extensive. A repeat study 1 week later shows evidence of a moderate amount of reperfusion (Fig. 3C).

Radionuclide images of other patients with pulmonary embolization are shown in Figures 4 through 7.

Underlying Pulmonary Disease

Areas of absent or reduced perfusion on the lung scan may be associated with any entity in which there is an interruption of pulmonary arterial blood flow. This is seen most commonly with pulmonary emphysema. Because of fibrotic changes involving the small vessels of the pulmonary arterial tree and destruction of alveoli, the pulmonary arterial blood flow may be markedly

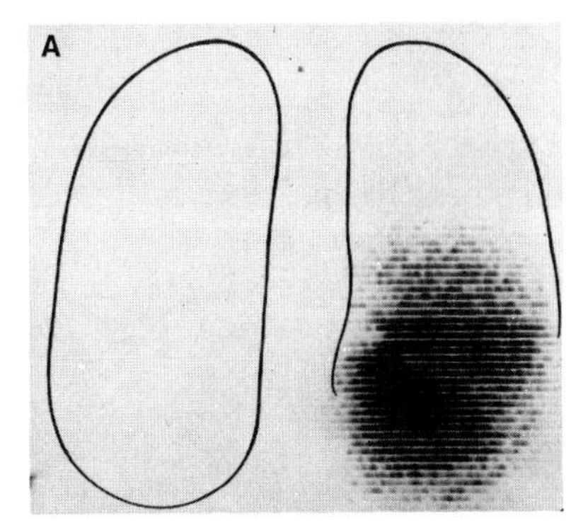

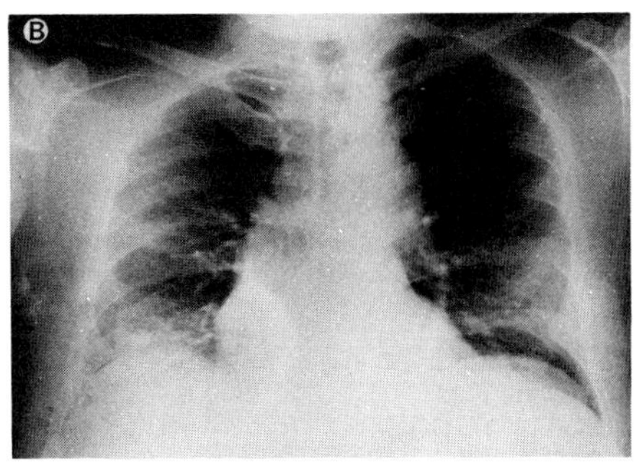

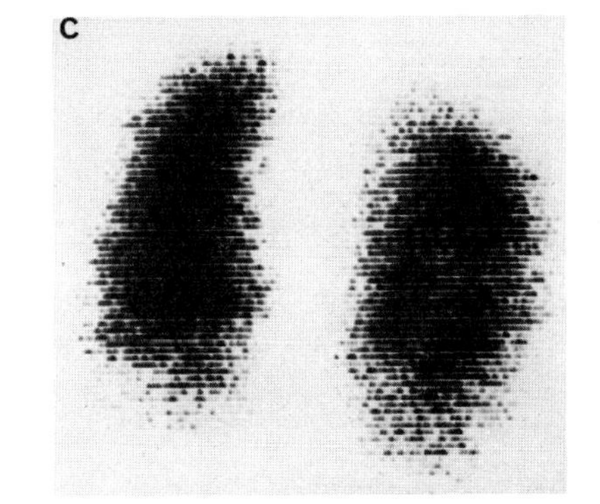

Fig. 3. A. Posterior lung scan in a patient with massive pulmonary embolism. There is extensive reduction in pulmonary arterial perfusion in all regions except the right lower lung field. The portions showing absent radioactivity are outlined. B. Chest x-ray at the time of the initial scan. C. Scan 2 weeks later shows marked improvement.

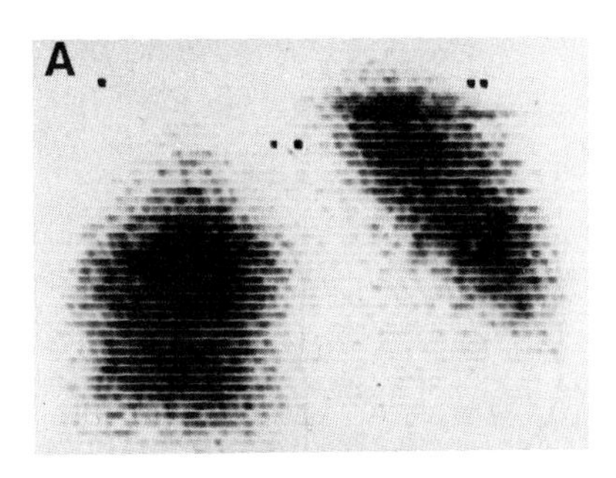

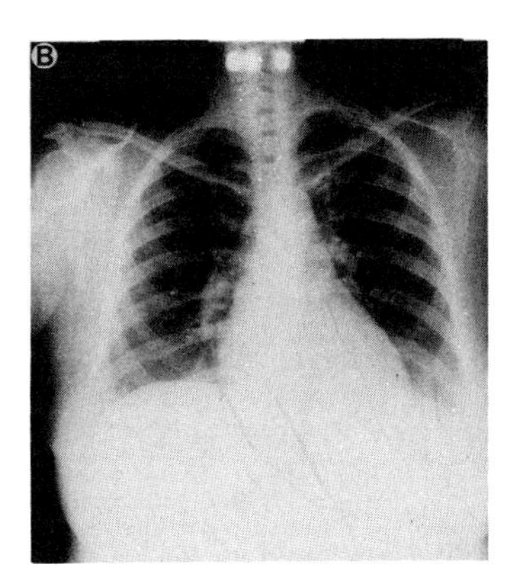

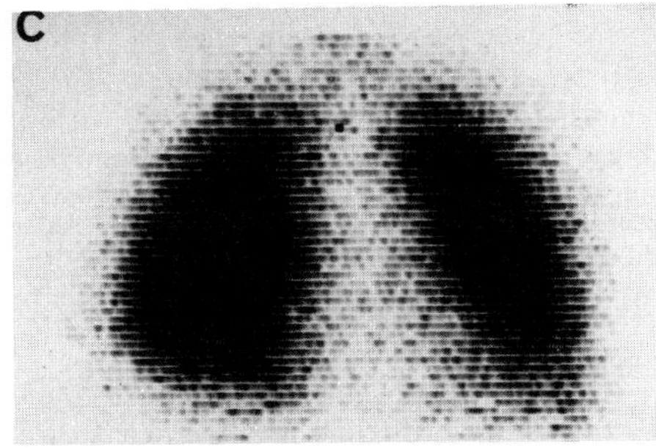

Fig. 4. A. Anterior scan of patient with pulmonary emboli showing areas of absent perfusion in right upper and left lower lung fields. B. Chest x-ray was normal. C. Scan 2 weeks later shows evidence of reperfusion. ^{131}I-MAA was used for scanning.

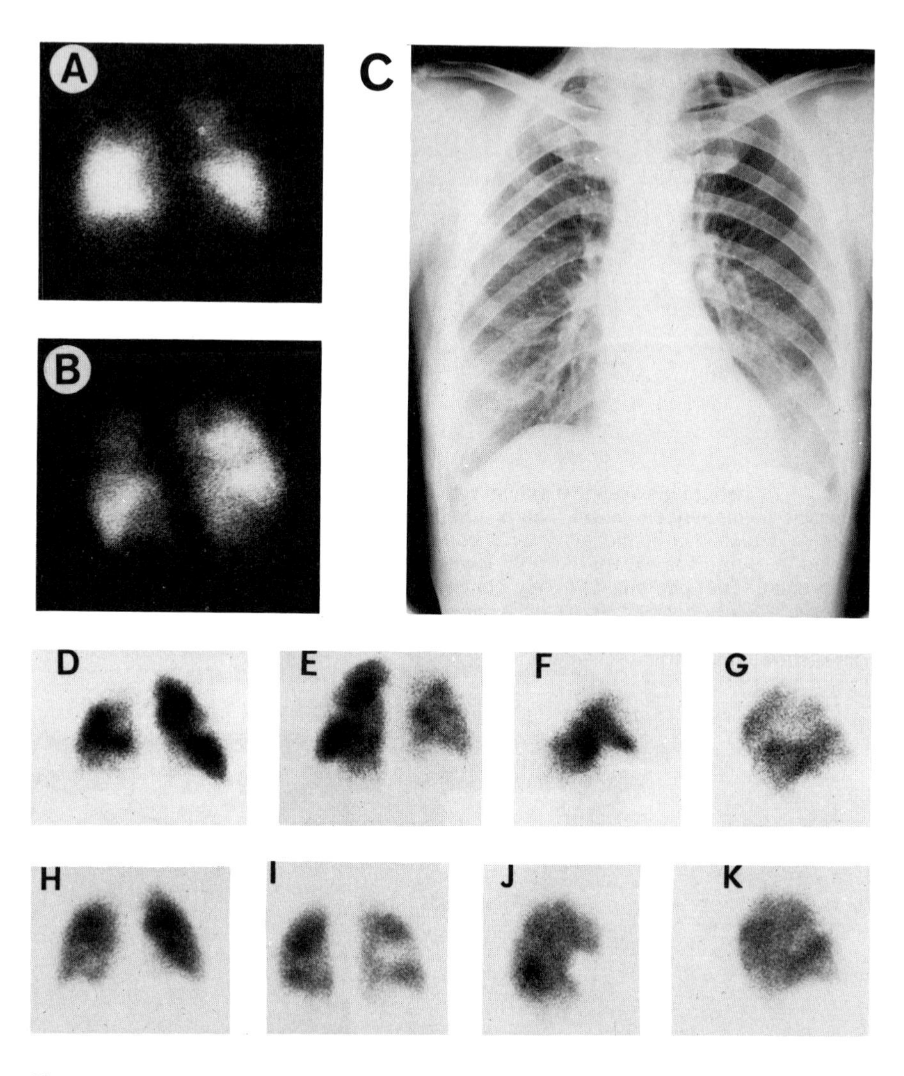

Fig.5. A, B, and C. *Multiple areas of pulmonary embolization.* Note the multiple sites of involvement in the anterior (A) and posterior (B) positions in the presence of a normal chest x-ray (C). D—K. *Multiple areas of pulmonary embolization with subsequent reperfusion.* The initial study shows many areas of absent perfusion in the anterior (D), posterior (E), right lateral (F), and left lateral (G) positions. The study 1 week later shows significant reperfusion and some evidence of further involvement as seen in the anterior (H), posterior (I), right lateral (J), and left lateral (K) positions. ^{99m}Tc-albumin microspheres were used for imaging.

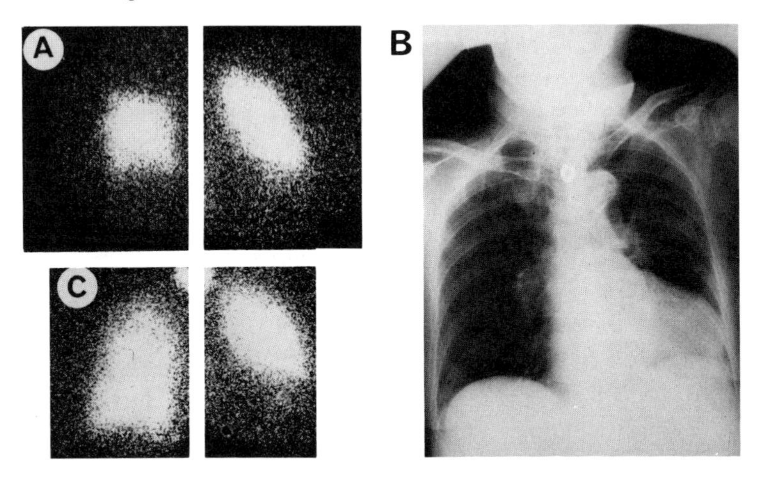

Fig. 6. Pulmonary imaging performed in the anterior position in a 77-year-old woman who developed anterior chest pain 1 week after surgery for a fractured hip. The large area of absent pulmonary arterial perfusion (A) in the absence of abnormalities on the chest x-ray in this region (B) makes pulmonary embolism the likely cause for the defect. C. Note the reperfusion 17 days later. ^{99m}Tc-albumin microspheres were used for imaging.

reduced to the involved areas. This manifests in varying degrees, depending on the severity of disease.

The areas of absent radioactivity demonstrated on such scans are not totally ischemic. They receive their arterial supply in large part from collaterals in the bronchial circulation. However, since the tagged particles are filtered out in the pulmonary capillaries, the presence of a collateral circulation to the involved areas cannot be demonstrated with radionuclide imaging.

It is common for perfusion defects to remain relatively constant over long periods of time in patients with chronic lung disease. The perfusion scan and chest x-ray of a patient with pulmonary emphysema are shown in Figures

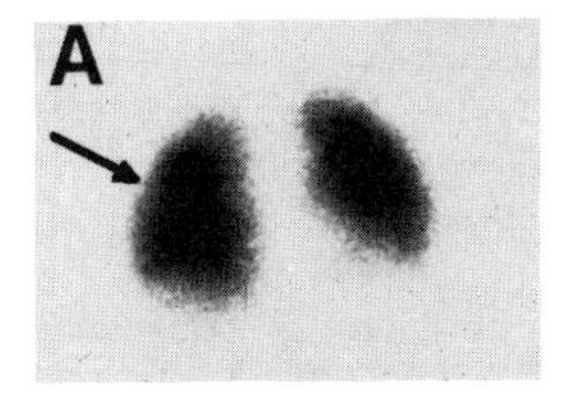

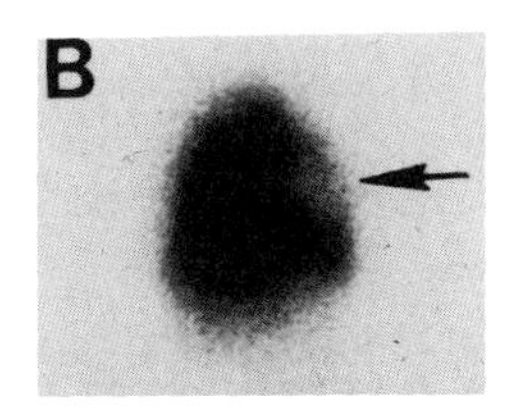

Fig. 7. Evidence of pulmonary embolization best seen is in the lateral position. Although there is barely a faint suggestion of reduced radioactivity (arrow) in the right upper lung field on the anterior image (A), the somewhat triangular area of absent pulmonary arterial perfusion is well seen in the right lateral position (B). ^{99m}Tc-albumin microspheres were used for scanning.

8A and B. Note the extensive areas of absent radioactivity on the scan. The scan of the same patient performed 3 months later showed no change. Perfusion images and accompanying chest x-ray of another patient with pulmonary emphysema are seen in Figures 8C and D.

The persistence in scan findings may help distinguish these patients from

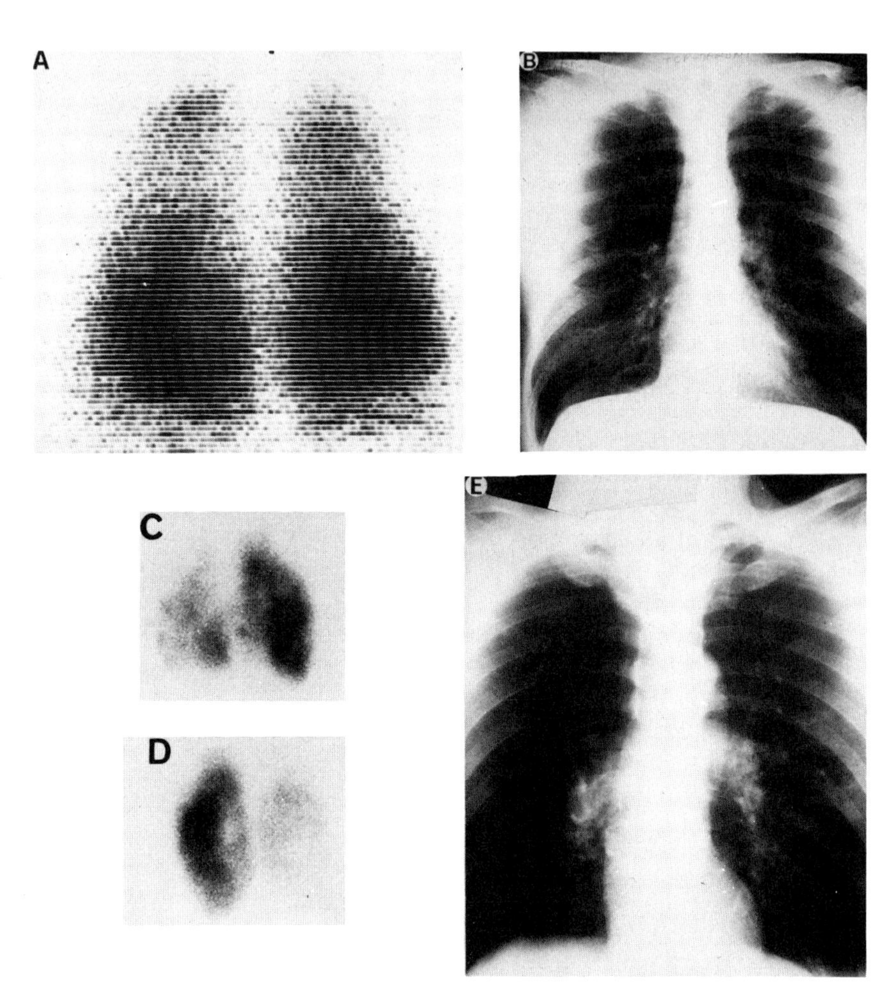

Fig. 8. Areas of absent pulmonary arterial perfusion in patients with pulmonary emphysema. *Reduced radioactivity in upper lung fields.* A. Posterior scan shows markedly reduced perfusion in both upper lung fields (using ^{131}I-MAA). B. The chest x-ray indicates uniform emphysematous changes. *Multiple areas of absent perfusion.* Imaging demonstrates numerous, irregular areas of absent perfusion in both the (C) anterior and (D) posterior positions, rather than the segmental areas of absent radioactivity so commonly seen with embolization (^{99m}Tc-albumin microspheres). E. The chest x-ray indicates the presence of chronic lung disease.

those with the rapidly changing patterns of pulmonary embolization. An exception is bronchial asthma, in which shunting of blood occurs during attacks, resulting in changing perfusion patterns.

Perfusion defects may be seen in a myriad of other pulmonary disease states. Pulmonary capillary thromboses associated with bacterial pneumonias are responsible for the absent perfusion seen in the involved segments (Fig. 9A and B). A tumor may cause external compression on an

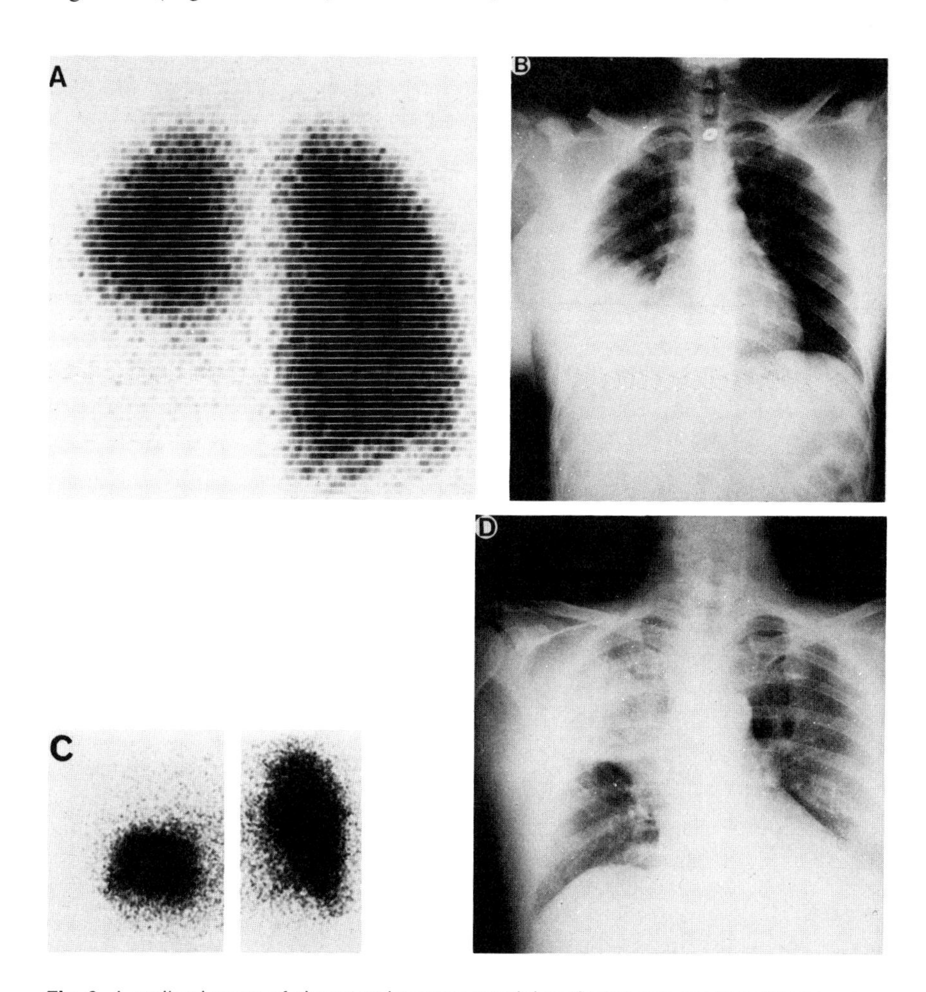

Fig. 9. Localized areas of absent pulmonary arterial perfusion not due to pulmonary embolization. A and B. Absent perfusion associated with lobar pneumonia. A.The large area of absent radioactivity in the right lower lung field seen after ^{131}I-MAA administration is secondary to capillary thromboses in a patient with pneumonia. B. Chest x-ray of same patient. C and D. Absent perfusion associated with neoplasm. C. Absent radioactivity from pulmonary arterial blockade due to bronchogenic carcinoma (^{99m}Tc-albumin microspheres). D. Chest x-ray of same patient.

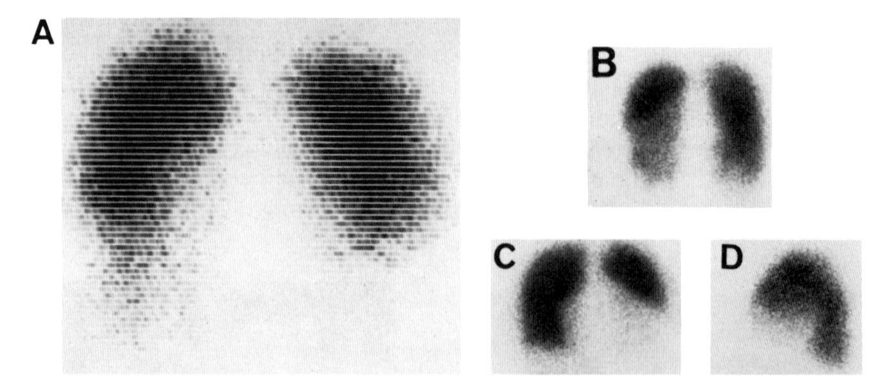

Fig. 10. A. Increased pulmonary arterial perfusion to the upper lung fields seen in the posterior position in A a patient with mitral stenosis (^{131}I-MAA). B. Increased pulmonary arterial perfusion to the upper lung fields in a patient with congestive heart failure and cardiomegaly, the latter evidenced in the anterior (C) and left lateral (D) positions. ^{99m}Tc-albumin microspheres were used for imaging.

adjacent pulmonary artery, resulting in arterial blockade and a perfusion defect on the scan (Figs. 9C and D).

On the other hand, there may be a regional increase in pulmonary arterial perfusion (Fig. 10) with a relative absence elsewhere in the lungs. This is apparent in mitral stenosis, in which increased perfusion to the upper lung fields is seen. It may be also detected in α_1-trypsin deficiency and in congestive heart failure (Fig. 10B).

RADIOACTIVE GASES

Radioactive gases are now commonly employed for pulmonary imaging and to obtain visual information about both the pulmonary arterial circulation and the respiratory tree itself.[6] The most frequently utilized gas is xenon 133, which may be used to obtain both perfusion and ventilation scans.

Perfusion Studies. For perfusion studies 20 to 30 mCi ^{133}Xe dissolved in approximately 1 ml saline is injected intravenously. Because ^{133}Xe is relatively insoluble in blood, when it reaches the pulmonary capillaries it is expired in the lungs. ^{133}Xe is delivered only to those portions of the lungs in which the pulmonary arterial circulation is intact. Thus an image of the lungs at the time the radioactive xenon reaches the alveoli reflects the state of pulmonary arterial perfusion. At the time of the procedure the patient is instructed to hold his breath for approximately 30 seconds following radionuclide administration. The scan images the radioactive material delivered to the lungs by the pulmonary arterial system alone. If there is a localized reduction in pulmonary arterial perfusion (e.g., due to an embolus), the radioactive gas does not enter those portions of the lungs and such regions appear as areas of absent radioactivity.

A limitation of perfusion scanning with a radioactive gas is that the scan can be performed in only one position and for no more than a brief moment after radionuclide administration. When the patient is no longer holding his breath the radioactive gas is distributed in a manner unrelated to the pulmonary vasculature.

Ventilation Studies. A ventilation scan may be performed by having the subject breathe and rebreathe a mixture containing ^{133}Xe in a closed system. This may be accomplished by rebreathing expired ^{133}Xe after its intravenous injection or by breathing the gas directly from a reservoir. The patient is instructed to inspire and while holding his breath a scan is performed over the next 15 to 30 seconds. The scan so obtained images the aerated portion of the respiratory tree. In normal inhalation imaging (Fig. 11A) there are no ob-

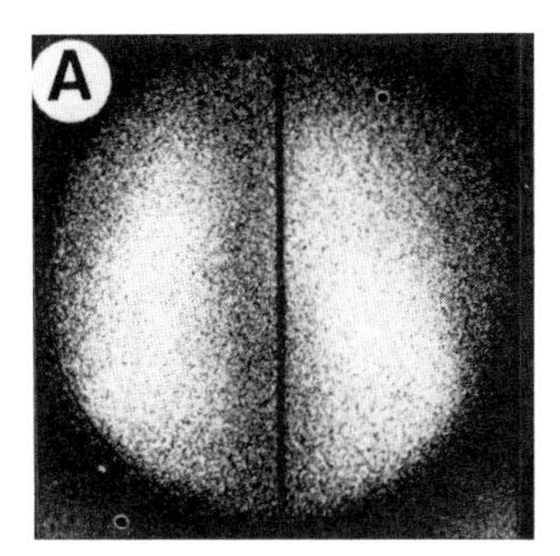

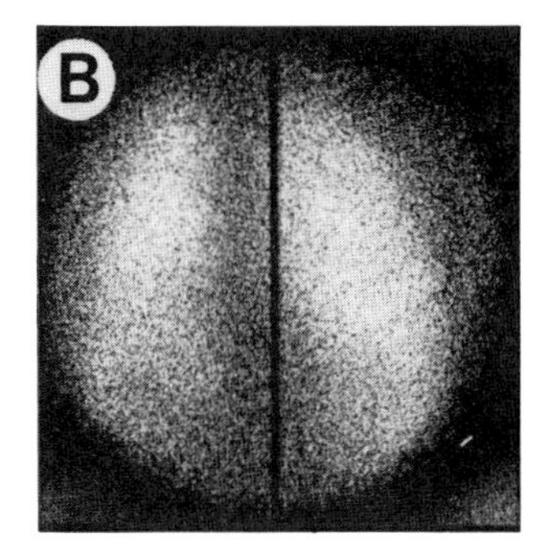

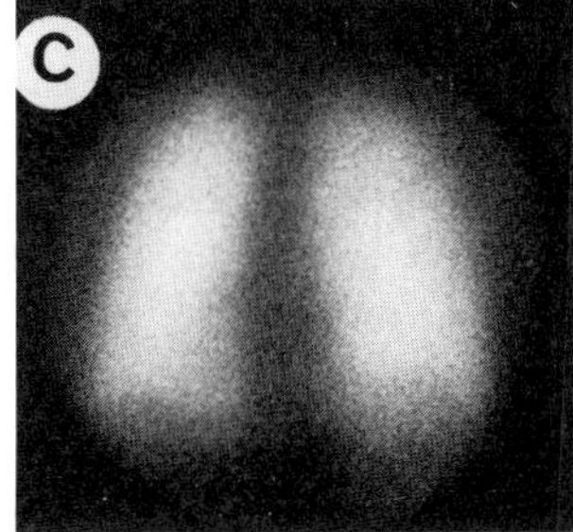

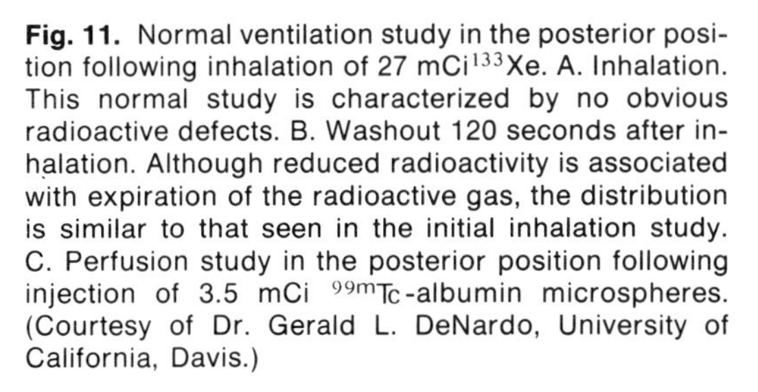

Fig. 11. Normal ventilation study in the posterior position following inhalation of 27 mCi ^{133}Xe. A. Inhalation. This normal study is characterized by no obvious radioactive defects. B. Washout 120 seconds after inhalation. Although reduced radioactivity is associated with expiration of the radioactive gas, the distribution is similar to that seen in the initial inhalation study. C. Perfusion study in the posterior position following injection of 3.5 mCi ^{99m}Tc-albumin microspheres. (Courtesy of Dr. Gerald L. DeNardo, University of California, Davis.)

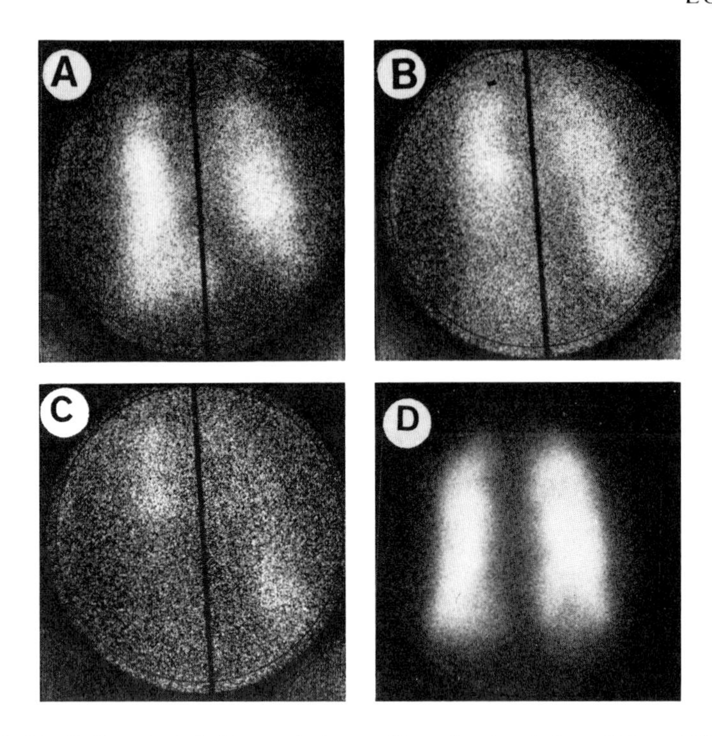

Fig. 12. Ventilation study in the posterior position after inhalation of 30 mCi ^{133}Xe in a patient with pulmonary emphysema. A. Inhalation. The most striking inhalational defect is in the region of the right lower lung field. B. Washout 120 seconds after inhalation. Note the relatively increased radioactivity in the left upper and right lower lung fields, associated with trapping of gas in these regions. C. Washout 240 seconds after inhalation. There is continued evidence of trapping in left upper and right lower lung fields. D. Perfusion study in same patient following injection of 3.5 mCi ^{99m}Tc-albumin microspheres. Note the difference between the perfusion defect at the right base and the extent of the inhalation defect in the same region in A. (Courtesy of Dr. Gerald L. DeNardo.)

vious areas of absent radioactivity. The patient is then allowed to breathe normally, and pulmonary imaging is then performed at suitable intervals, such as every 10 to 30 seconds, over the next three or four minutes. This expiratory or washout study images that portion of the gas that remains in the respiratory passages (Fig. 11B).

The inhalation scan of a patient with pulmonary emphysema is seen in Figure 12A. The areas of absent radioactivity indicate regions not penetrated by the radioactive gas and correspond to those portions of the lung not undergoing aeration. Thus the functional portions of the respiratory tree may be imaged in a manner unknown before the use of radioactive materials. The washout or expiratory scan shows areas of radioactivity remaining in the

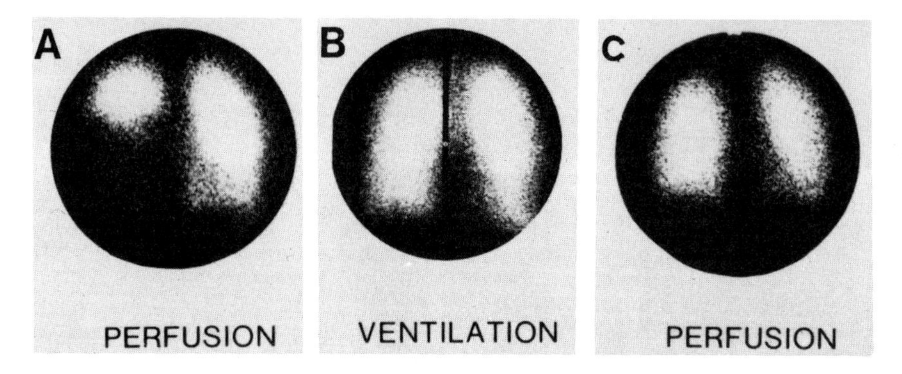

Fig. 13. Ventilation and perfusion imaging with pulmonary embolization. A. Initial ^{131}I-MAA perfusion image performed in the anterior position shows absent radioactivity in the right lower lung field in the area of embolization. B. Ventilation study performed with ^{133}Xe is normal. C. Perfusion imaging performed 9 days later is now normal and consistent with reperfusion. (From DeNardo et al: N Engl J Med 282:1334, 1970).

respiratory tree (Fig. 12B). This corresponds to the air trapped in the respiratory passages of this patient.

The ^{133}Xe ventilation scan has proved a useful adjunct in the diagnosis of pulmonary embolism. Although pulmonary arterial perfusion is impaired, ventilation may remain normal in the involved regions[5,7] (Fig. 13).

RADIOACTIVE AEROSOLS

Another less frequently employed method for performing ventilation imaging is through inhalation of a radioactive aerosol.[2] The patient breathes this material in a closed system, and the aerosolized radioactive solution is delivered to the respiratory tree through either positive pressure or with an ultrasonic nebulizer. The radioactive particles range from 0.1 to 2.0μ in diameter. About 10 percent of these inhaled particles reach the alveoli, where they are desposited on the epithelial surfaces that are being aerated. Like the ^{133}Xe ventilation scan, the aerosol inhalation study depicts those portions of the lung actively participating in respiration.

Although a large number of radioactive materials have been employed in aerosol scanning, ^{99m}Tc-labeled albumin is most commonly used. A satisfactory nebulizing solution contains 3 mCi in 7 ml.

REFERENCES

1. Wagner HN Jr, Sabiston DC Jr, McAfee JG: Diagnosis of massive pulmonary embolism in man by radioisotope scanning. N Engl J Med 271: 377, 1964
2. Taplin GV, Dore EK, Poe ND, et al: Pulmonary arterial perfusion and aerated space assessment by scintiscanning. In Simon M, Potchen EJ, LeMay M (eds): Frontiers in Pulmonary Radiology. New York, Grune & Stratton, 1969, pp 33-75
3. Galt JM, Tothill P: The fate and dosimetry of two lung scanning agents, ^{131}I-MAA and ^{99m}Tc ferrous hydroxide. Br J Radiol 46:272, 1973
4. Tow DW, Wagner HN Jr: Recovery of pulmonary arterial blood flow in patients with pulmonary embolism. N Engl J Med 276:1053, 1967
5. De Nardo GL, Goodwin DA, Ravasini R, et al: The ventilatory lung scan in the diagnosis of pulmonary embolism. N Engl J Med 282:1334, 1970
6. West JB: Distribution of blood and gas in the lungs. Phys Med Biol 11:357, 1966
7. McNeil BJ, Holman BL, Adelstein SJ: The scintigraphic definition of pulmonary embolism. JAMA 227: 753, 1974.

ADDITIONAL READING

DeLand FH, Wagner HN Jr: Lung and Heart, vol II. In Atlas of Nuclear Medicine. Philadelphia, Saunders, 1970, pp 1-215.

Gilson AJ, Smoak WM III (eds): Pulmonary Investigation with Radionuclides. First Annual Nuclear Medicine Seminar. Springfield, Charles C. Thomas, 1970

Holman BL, Lindeman JF (eds): Progress in Nuclear Medicine: Regional Pulmonary Function in Health and Disease. Baltimore, University Park Press, 1973

Poe ND, Taplin GV: Pulmonary scanning. In Blahd WH (ed): Nuclear Medicine, 2nd ed. New York, McGraw-Hill, 1971, pp 322-349.

Mishkin FS, Brashear RE: Use and Interpretation of the Lung Scan. Springfield, Charles C. Thomas, 1971

Wagner HN Jr, Holmes RA, Lopez-Majano V, et al: The lung. In Wagner HN Jr (ed): Principles of Nuclear Medicine. Philadelphia, Saunders, 1968, pp 472-530

chapter 6

OTHER IMAGING AND MONITORING PROCEDURES

BONE IMAGING

Before neoplastic involvement of bone can be seen on a roentgenogram, at least 60 percent of the affected area of bone must have undergone destruction or replacement by tumor.[1] It is thus surprising that there may be extensive tumor involvement sufficient to give painful symptoms without x-ray evidence of the lesion. On the other hand, evidence of tumor can be detected on bone scans at a far earlier stage of neoplastic growth, and this successful demonstration has been largely responsible for the prominence radionuclide bone imaging has attained in recent years.

Although there are notable exceptions, the usual response of bone to the presence of tumor is to attempt to form new bone at the site or in the periphery of the tumor. This is commonly referred to as reactive bone formation and occurs with both osteolytic and osteoblastic lesions. Bone mineral appears as hydroxyapatite, $3Ca_3(PO_4)_2 \bullet Ca(OH)_2$, in the form of crystals arranged in a lattice work.

An objective, then, of bone scanning is to provide a visual representation of radionuclide accumulation in crystal growth or accretion. Isotopes of calcium (e.g., ^{47}Ca and ^{49}Ca) cannot be used for this purpose because their energies are too high for scanning. Therefore other radioactive materials must be employed that have suitable energies and that can freely exchange with one of the ions in the hydroxyapatite crystal. The isotopes most commonly used are ^{85}Sr, ^{87m}Sr, ^{18}F, and ^{99m}Tc. Delivery of these materials is of course dependent on the arterial supply to the areas in question. One must bear in mind too that there are countless situations other than tumor involvement in which increased amounts of hydroxyapatite collect (e.g., fracture, callus formation).

Strontium 85 Studies

Insofar as bone crystal is concerned, the metabolism of strontium is similar to that of calcium. Thus strontium ions freely exchange with and replace calcium ions in bone crystal.

^{85}Sr was the first isotope to be widely employed for skeletal imaging and has been used extensively during the past decade.[2,3] It is still the isotope of choice for bone scanning in many centers. The usual dose is 100 to 125 uCi administered intravenously, and is available as either ^{85}Sr-strontium chloride or ^{85}Sr-strontium nitrate. Since it has a relatively long half-life (64 days) and a principal gamma emission of 510 keV, the radiation exposure is appreciably greater than with the other more recently developed radioactive materials.

Scanning is not attempted until at least 48 hours and preferably 72 hours after radionuclide injection. Although a considerable quantity of the tracer enters bone during the first 2 to 3 hours, the blood level of radioactivity then is so high that scanning is not possible. In the interim there is continued skeletal deposition of ^{85}Sr, and the unfixed isotope is excreted through the kidneys and into the intestinal tract. When imaging of structures that overlie the intestinal tract is desired (e.g., lumbosacral spine and pelvis) it is necessary to give a cleansing enema 2 to 3 hours before scanning because of intestinal excretion of the ^{85}Sr.

Since there is a constant renewal of elements in bone crystal, the tracers used in bone scanning enter the entire skeleton. However, increased radioactivity is seen at sites in which excessive crystal formation has occurred, as with neoplasms of bone.

A normal bone scan is seen in Figure 1. Note the uniform deposition of the radioactive material in the vertebrae. The relatively increased radioactivity in the regions of joints is a reflection of the increased calcium turnover at articular surfaces.

The bone scan of a patient with carcinoma of the breast hospitalized because of low back pain shows a localized area of increased radioactivity (Fig. 2), which is consistent with neoplastic involvement in this region. The corresponding roentgenogram shows no such abnormality. Note that the areas on the scan adjacent to the lesion show little or no radioactivity. Since calibration was made on the localized area of increased radioactivity, the surrounding regions of normal bone in which the ^{85}Sr uptake is considerably less was suppressed in setting up the scan.

This localized area of increased radioactivity in the first lumbar

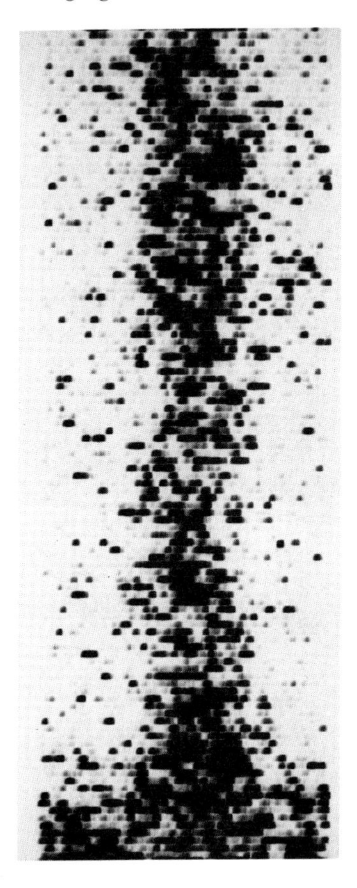

Fig. 1. Normal scan of the lower dorsal and lumbar vertebrae performed in the posterior position 72 hours after ^{85}Sr-strontium nitrate injection.

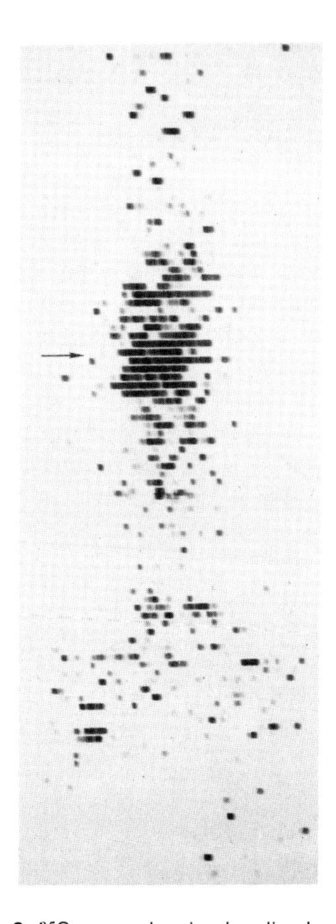

Fig. 2. ^{85}Sr scan showing localized area of increased radioactivity (arrow) in lumbar vertebrae indicating site of metastatic involvement.

vertebra, however, does not establish that neoplastic involvement of bone is present in this region. It simply means that there is increased strontium deposition here, and in the absence of any pathologic change on the x-rays neoplasm is most likely. ^{85}Sr scanning showing neoplastic involvement of other bones is seen in Figures 3 and 4. The findings are similar for both primary and metastatic tumors.

Patients often experience remarkable symptomatic relief of pain with radiation therapy when bony metastases can be demonstrated on scans but insufficient decalcification has occurred for roentgenologic visualization. This is probably the single most important factor for pursuing the search for bony metastases in the patient with a known primary neoplasm.

Countless other situations exist in which there is bone crystal accumulation and of course an associated increase in ^{85}Sr deposition. These include, among others, fracture, callus formation, osteomyelitis, and Paget's disease. In some instances more than one disorder associated with increased ^{85}Sr uptake is present. This is commonly encountered with patients suspected of having a pathologic fracture in whom neoplastic involvement of bone cannot be detected on roentgenograms. Because ^{85}Sr depositions in the areas of both fracture and neoplasm are high and similar, bone scanning here is unrewarding for detecting tumor at the fracture site.

However, in some situations the difference in ^{85}Sr uptake in the two entities may be of sufficient magnitude so that each may be imaged separately. This is not uncommonly seen in an area of sarcomatous degeneration in a patient with Paget's disease. The deposition of ^{85}Sr associated with the neoplasm may exceed that due to the underlying disease, so the site of the sarcoma may be clearly defined on the bone scan.

There are several abnormalities not always associated with reactive bone formation: highly anaplastic tumors, old osteoblastic lesions that are no

Fig. 3. Areas of increased radioactivity after ^{85}Sr administration showing bone metastases in pelvis and lumbar spine. The primary tumor was adenocarcinoma of the breast.

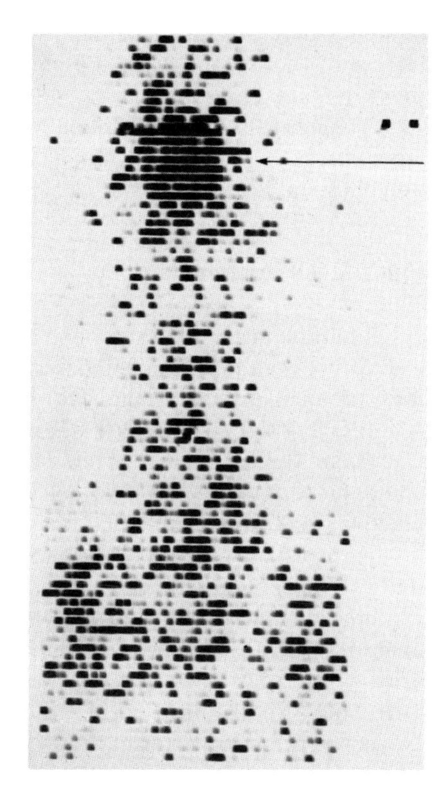

Fig. 4. ^{85}Sr scan performed in the posterior position in a patient with bronchogenic carcinoma. Metastatic involvement is indicated by the localized area of increased radioactivity (arrow) in the region of the twelfth dorsal vertebra.

longer expanding, and certain "punched out" lesions such as are seen with multiple myeloma and the histiocytoses. Because there may be no unusual accretion of bone crystal in these conditions there is no related increased uptake of ^{85}Sr.

Strontium 87m Studies

Bone scanning with ionic ^{87m}Sr has been disappointing in many respects. The isotope has a short half-life (2.8 hours) and a principal gamma emission of 390 keV. The mechanism of uptake in bone is identical to that of ^{85}Sr.[2] Scanning may be performed 1 hour after isotope injection.

Although a considerable amount of radioactive material is deposited in bone shortly after tracer administration, the blood level of ^{87m}Sr remains so high that the scanning procedure may be technically difficult. In many instances in which there has been neoplastic involvement of bone, the accumulation of radioactive strontium in the bone crystal produced in response to tumor occurs gradually over a period of days. Since the half-life of ^{87m}Sr is so short, there is little opportunity to detect a bony lesion 3 to 4 days after injection of the isotope, with subsequent failure to image neoplastic involvement in some cases.

The need for a suitable bone scanning agent with a short half-life has in large part been satisfied by the development of ^{18}F-sodium fluoride and ^{99m}Tc-polyphosphates.

Fluorine 18 Studies

Fluoride accumulates in bone crystal too, but the mechansim of uptake is different than that seen with strontium.[4] Fluoride exchanges with hydroxyl ions on the surface of the hydroxyapatite crystal to form fluorapatite, $3Ca_3(PO_4)_2 \bullet CaF_2$. Because the concentration of hydroxyl ions in extracellular fluid is relatively low, fluoride may enter bone crystal promptly and in large amounts. Thus ^{18}F uptake may be a more sensitive barometer of reactive bone formation, whatever the cause, than radioactive strontium.

^{18}F is usually cyclotron-produced, and its use is dependent on the proximity to a supplier with such a facility. Because it is a positron emitter, the annihilation photon with a gamma emission of 510 keV may be used for scanning. It has a relatively short half-life and is excreted principally via the urinary tract. Scanning may be started 2 hours after intravenous injection of 1 to 2 mCi ^{18}F-sodium fluoride. It is important that the urinary bladder be evacuated prior to performance of the scan.

A bone scan performed following ^{18}F injection is shown in Figure 5. The areas of increased radioactivity, indicating neoplastic involvement, are similar in appearance to those observed in radioactive strontium studies.

In addition, ^{18}F scans may be useful in visualizing areas of reactive bone in some patients with nonneoplastic lesions whose roentgenograms do not yet show evidence of bone destruction or repair (e.g., in cases of aseptic necrosis in which there is reactive bone formation adjacent to the sites of necrosis). This is due to increased hydroxyapatite crystal deposition in response to the necrotic changes; subsequent roentgenograms usually show the area of aseptic necrosis not previously visualized.

^{99}. Tc-Polyphosphate Studies

The success in demonstrating bone tumors with radiostrontium and radiofluoride was merely a prelude to another exciting advance in nuclear medicine: bone imaging with ^{99m}Tc-polyphosphates. Developed by Subramanian and his associates, these compounds have been used to image the skeleton in remarkable detail.[5] Such imaging was hardly anticipated even as recently as 5 years ago.

The mechanism of uptake of the polyphosphates has not yet been clearly defined and may very well be associated with both calcium and hydroxyl ion

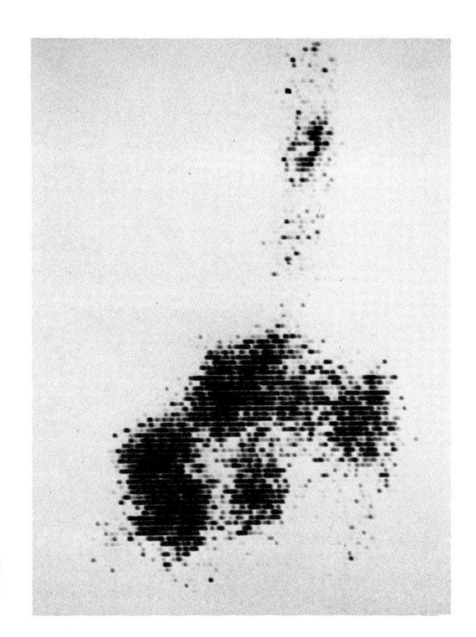

Fig. 5. Posterior scan showing pelvic metastases 2 hours after ^{18}F-sodium fluoride administration.

replacement in hydroxyapatite crystal. The usual adult dose of this radionuclide is 5 to 10 mCi, and imaging may commence 1 to 2 hours after tracer injection. An important factor in the enthusiasm for ^{99m}Tc-polyphosphates is that the skeleton can be imaged successfully in a relatively short period of time with a scintillation camera rather than a rectilinear scanner. Because the radioactive material is excreted through the kidneys there is often excellent visualization of the structures in the urinary tract; it is not unusual, for example, to have a case of hydronephrosis become manifest on the ^{99m}Tc-polyphosphate image.

Normal ^{99m}Tc-polyphosphate imaging is seen in Figure 6. Note the detail with which the ribs and vertebrae are seen. Bone metastases are very clearly visualized with these studies (Figs. 7 through 9). Note the clarity with which the small area of involvement in the third lumbar vertebra is seen in Figure 7.

More recently, ^{99m}Tc-labeled 1-hydroxyethylidene-1, 1-disodium phosphonate (^{99m}Tc-diphosphonate) has been successfully employed for skeletal imaging.[5a,5b] Its development followed the observation that diphosphonates employed clinically, as in Paget's disease, become chemisorbed to hydroxyapatite crystal to prevent the latter from undergoing accretion or dissolution.

Especially since the introduction of ^{99m}Tc bone imaging, radionuclide studies are frequently employed to help determine more precisely the extent of involvement of a nonneoplastic process that was originally detected on roentgenograms (e.g., osteomyelitis and Paget's disease). Imaging of the latter problem is seen in Figure 10. In this case the ^{99m}Tc study shows that the involvement is confined to a single vertebra, just as the x-rays demonstrated.

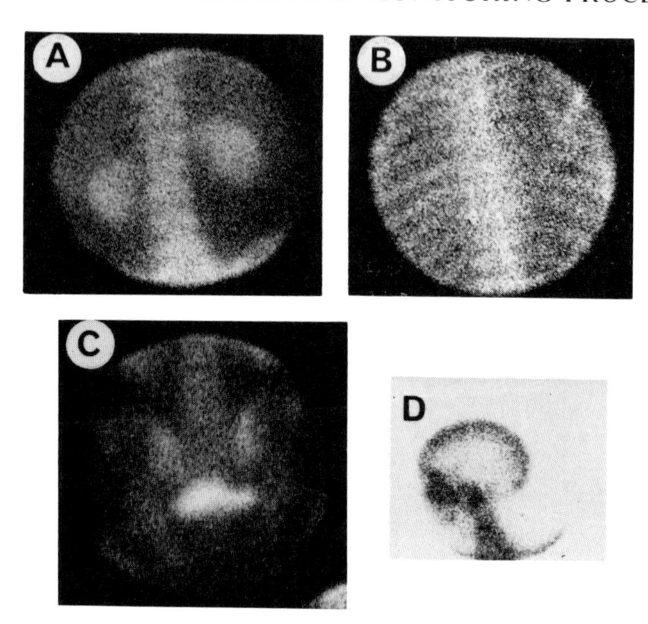

Fig. 6. Normal skeletal studies 2 hours following ^{99m}Tc-stannous polyphosphate injection. Normal vertebral (A), costal (B), and pelvic (C) imaging, all performed in the posterior position. (D). Normal imaging of the skull in the left lateral position.

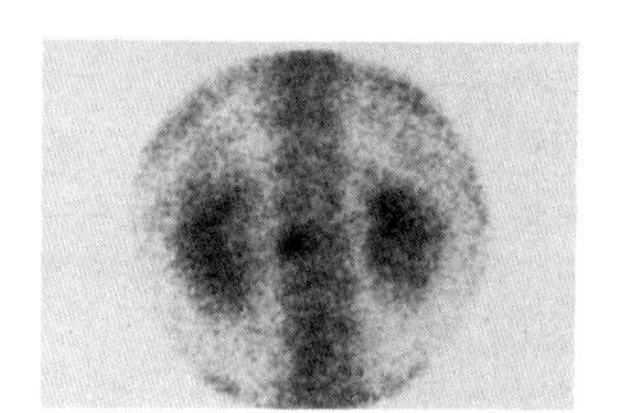

Fig. 7. Metastatic carcinoma of the breast with involvement of the third lumbar vertebra demonstrated after injection of ^{99m}Tc-polyphosphate. The small area of increased activity indicates the site of involvement.

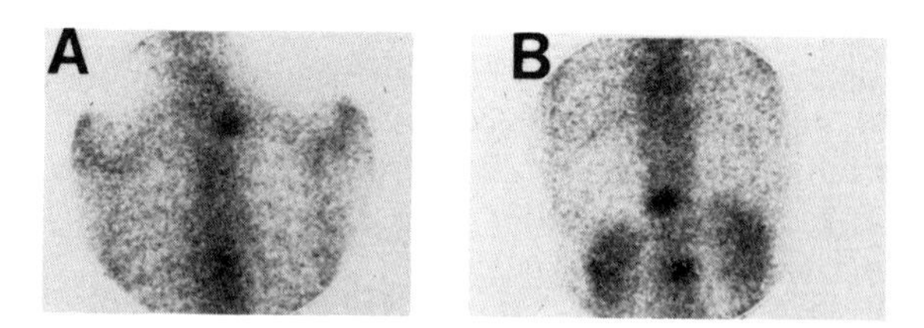

Fig. 8. ^{99m}Tc-stannous polyphosphate study showing localized areas of increased radioactivity in the second dorsal (A) and second and fourth lumbar (B) vertebrae due to metastatic involvement from bronchogenic carcinoma. Note how easily the anatomic sites of the involved areas are determined from the radionuclide image.

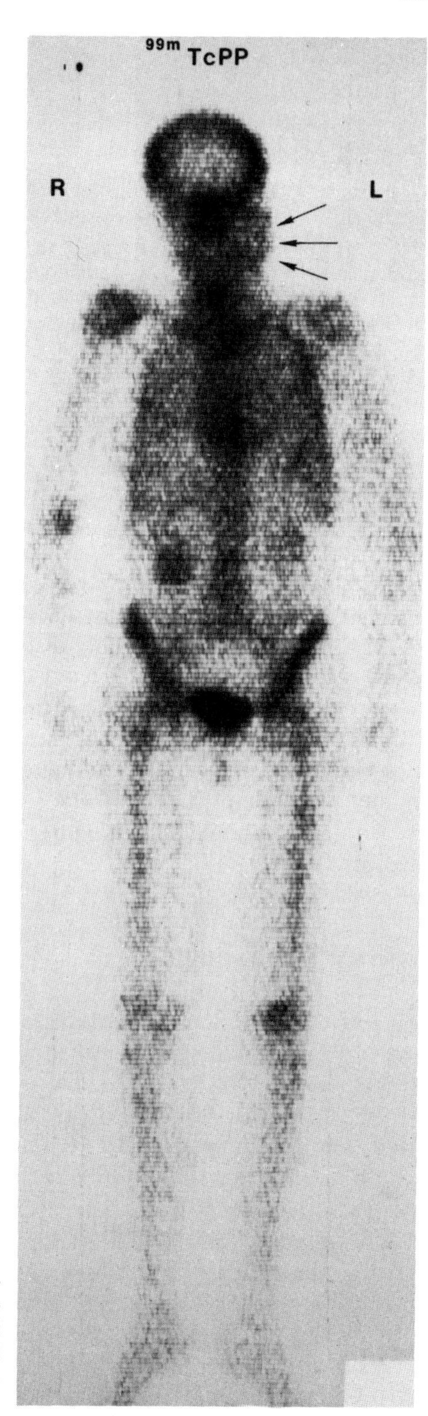

Fig. 9. Skeletal scan in the anterior position after injection of ^{99m}Tc-polyphosphate. Extension of squamous cell carcinoma to the left mandible is indicated by arrows. (Courtesy of Dr. Paul A. Farrer, Royal Victoria Hospital, Montreal.)

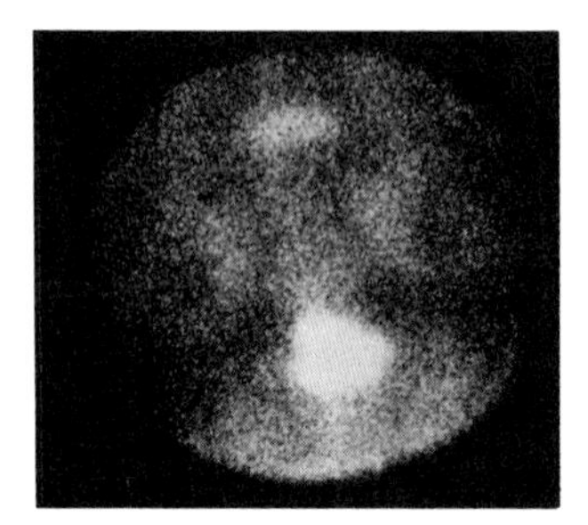

Fig. 10. Imaging of Paget's disease with ^{99m}Tc-stannous polyphosphate. The area of involvement is confined to L3.

PLACENTOGRAPHY

Radionuclide studies have come to play an important role in diagnosing placenta previa during the past decade. A radioactive material such as ^{131}I-HSA, which becomes uniformly distributed in the circulatory system, is injected intravenously. The placenta can then be located, either by external monitoring or imaging, because of the relatively high radioactivity associated with the entry of the radionuclide into the intervillous space of the placenta. It is essential that the radioactive material remain in the maternal circulation and not cross the placental barrier to enter the fetal circulation lest there be undesirable radionuclide accumulation in the fetal organs (e.g., the fetal thyroid).

External Monitoring

For external monitoring[6] the uterus is palpated and its projection to the anterior abdominal wall identified and outlined with a marking pen. The uterine area is further divided into nine segments (Fig. 11). Following pretreatment with Lugol's to prevent ^{131}I uptake in the maternal thyroid gland, the patient is given 10 μCi ^{131}I-HSA intravenously; after waiting 15 minutes to ensure adequate mixing of the radionuclide, external monitoring over each outlined segment and over the heart is commenced. Counts are usually obtained for a period of 1 minute in each region.

If a placenta previa is present, increased radioactivity is observed in the lower three segments, depending on whether the placenta previa is total, partial, or marginal. The radioactivity at the site of placental implantation approaches that present in the circulatory system, as manifested by the radioac-

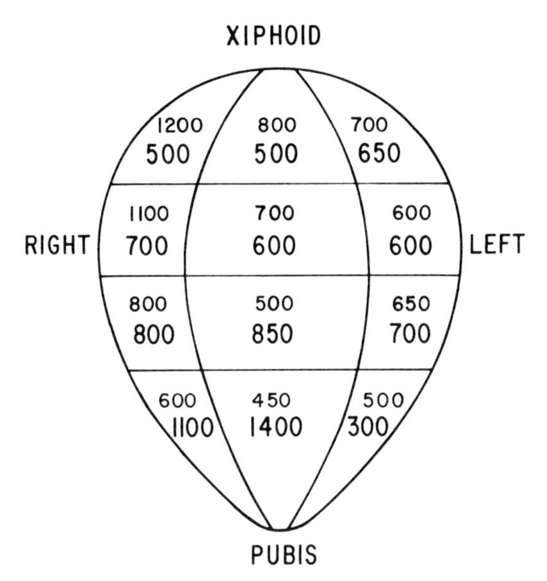

Fig. 11. Segments used in external monitoring for possible placenta previa. The small (upper) numbers indicate the counts per minute in a patient with no evidence of placenta previa; the larger (lower) numbers show the radioactive counts in a patient with placenta previa.

tive counts obtained over the heart. Normal implantation of the placenta is indicated by the greatest radioactivity being in one of the upper six segments. Care must be taken that radioactivity in the heart and liver is not reflected in the counts over the uterus.

Statistics as to the accuracy of diagnosing placental position by external monitoring vary among institutions, but in diagnosing placenta previa it is probably upward of 95 percent. The low count rate obtained after only 1 minute of monitoring over each segment may be a source of error.

Placental Imaging

Among the materials used successfully for placental imaging[7] are ^{99m}Tc-HSA and ^{113m}In-transferrin. Lateral and anterior images of a patient with a normally implanted placenta are seen in Figure 12. In this case implantation in the left anterolateral fundus is readily discerned. Imaging of the upper abdomen shows evidence of increased radioactivity toward the left (Fig. 12B), while imaging in the left lateral position (Fig. 12C) demonstrates the anterior position of the implantation. Note the failure to visualize the placental blood pool in the image of the lower abdomen (Fig. 12A).

Placenta previa is readily demonstrated on the radionuclide image. Imaging of the lower abdomen (Fig. 13A) in a patient with a central placenta previa shows increased radioactivity centrally in the suprapubic region; in the right lateral position (Fig. 13B) the placental implantation overlies the region

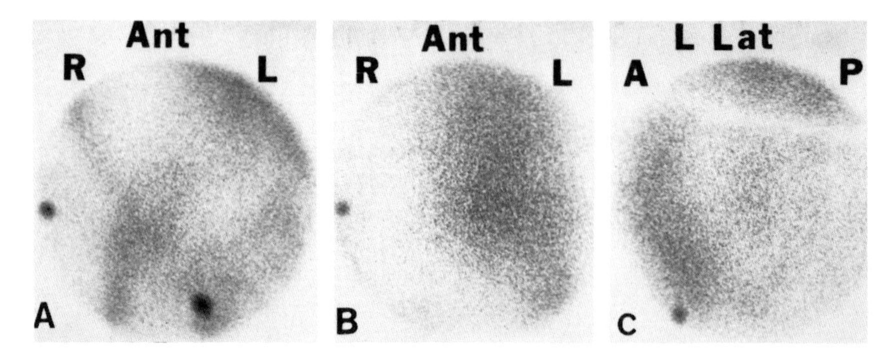

Fig. 12. Normal placental imaging following administration of 1 mCi ^{99m}Tc-HSA in a patient with left anterolateral placental implantation. A. Anterior position, lower abdomen. There is no evidence of increased radioactivity associated with the placental blood pool. Markers are on symphysis pubis and right lateral body wall. B. Anterior position, upper abdomen. Placental blood pool is easily seen toward the left. C. Left lateral position. The anterior implantation is readily identified. Note the position of the left costal margin (arrow). (Courtesy of Dr. Ralph J Gorten, University of Texas, Galveston.)

of the internal os uterus and extends a bit more anteriorly than posteriorly.

Imaging in cases of placenta previa has obvious advantages over external monitoring, and accuracy of the two methods is comparable. Roentgenologic diagnosis of placenta previa for the most part has been replaced by ultrasound and radionuclide studies.

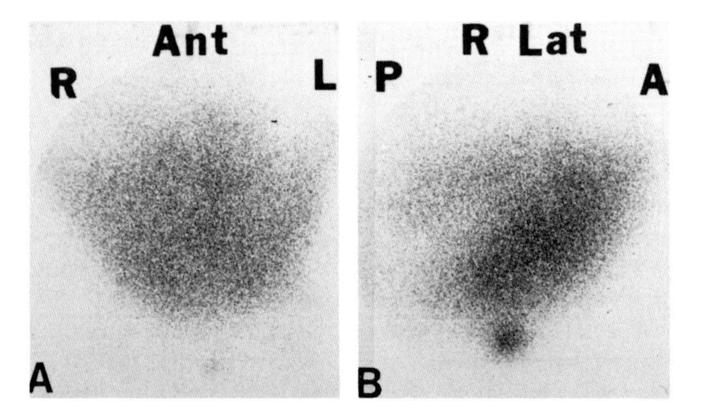

Fig. 13. Central placenta previa demonstrated after injection of 1 mCi ^{99m}Tc-HSA. A. Anterior position, lower abdomen. Note increased radioactivity in the suprapubic area due to the placental blood pool. B. Right lateral position. The site of implantation overlies the region of the internal os uterus and extends a bit more anteriorly than posteriorly. Note the site of the implantation in relation to the marker over the symphysis pubis. (Courtesy of Dr. Ralph J. Gorten, University of Texas, Galveston.)

IMAGING FOR PERICARDIAL EFFUSION

The availability of radionuclide methods to image the cardiac blood pool has been applied to the search for pericardial effusions.[8] The radioactive materials used most successfully in this area are those that remain confined to the circulatory system after intravenous administration (e.g., ^{99m}Tc-HSA and ^{131}I-HSA). Currently technetium 99m is the radionuclide of choice, with a usual adult dose of 3 mCi.

Cardiac imaging is performed in the anterior position when using a rectilinear scanner. The scan is superimposed on a chest x-ray, and normally there is no disparity between the image of the cardiac vascular pool and the cardiac silhouette of the chest x-ray (Fig. 14A). In order to minimize distortion an x-ray film obtained with a split-film technique is helpful. The hepatic image is usually clearly visualized as the radioactive material perfuses through the liver and is closely related to that of the cardiac vascular pool. Although the chest x-ray may sometimes suggest the possibility of pericardial effusion in a patient with cardiac dilatation, failure to demonstrate a cardiac silhouette-tracer blood pool disparity can rule out a diagnosis of effusion (Fig. 14B).

If a pericardial effusion is present, there is usually a noticeable disparity between the vascular pool scan and the cardiac silhouette of the chest x-ray (Fig. 15). This disparity represents the region of the effusion. Before it can be appreciated with scanning techniques the pericardial effusion must contain at least 100 ml fluid.

In most cases of pericardial effusion the hepatic image becomes separated from the cardiac vascular pool to varying degrees because of imposition of the effusion between heart and liver. However, such separation is an inconstant finding, and in some instances no separation is detected when a pericardial effusion is present. This is especially true when a loculated effusion has formed.

When using a scintillation camera the presence or absence of a pericardial effusion may be determined by the degree of separation between the images of the cardiac vascular pool and pulmonary vasculature.[9] A normal study is seen in Figure 16. The radionuclide image of a patient with a large pericardial effusion (Fig. 17) demonstrates a wide separation between the cardiac and pulmonary vascular images.

It is sometimes helpful to utilize rapid sequence imaging of the cardiac area immediately after radionuclide injection. This may more clearly define

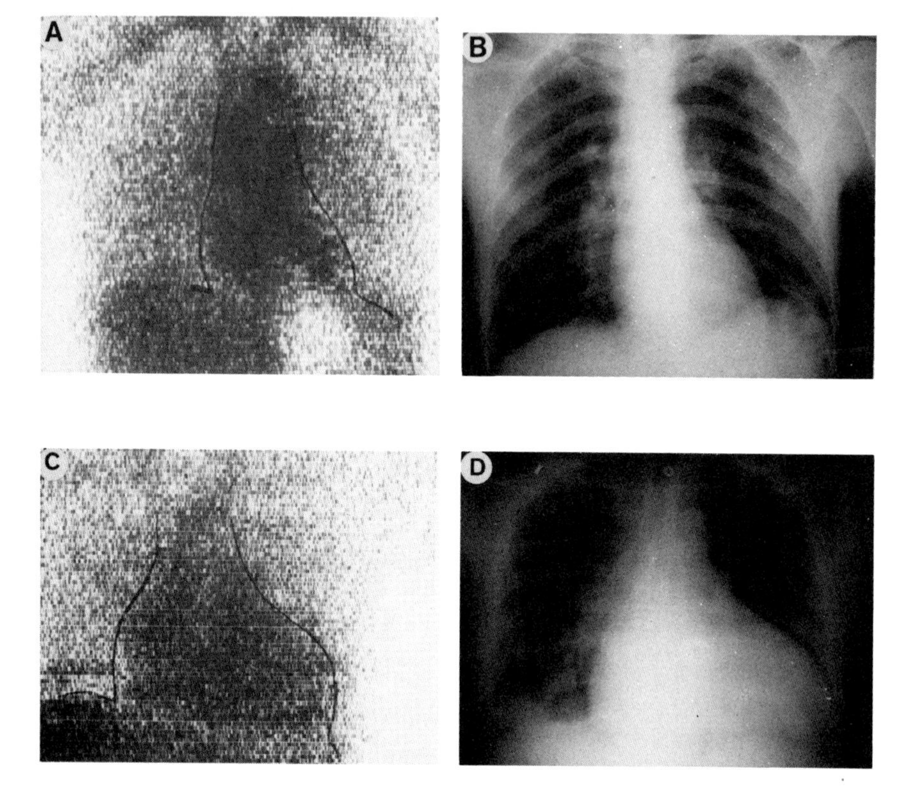

Fig. 14. Absence of pericardial effusion. A and B. Normal. The rectilinear scan performed after ^{99m}Tc-HSA administration shows no disparity between the cardiac vascular pool (A) and the cardiac silhouette of the chest x-ray (B), the latter outlined on the scan. C and D. Cardiac dilatation. Although the heart is enlarged, the absence of a pericardial effusion is indicated by the lack of disparity between the cardiac vascular pool seen on radionuclide imaging (C) and the cardiac silhouette (D). (Courtesy of Dr. Ralph J. Gorten, University of Texas, Galveston.)

any possible radionuclide separation between pulmonary and cardiac blood pools (see "Radionuclide Angiography" in this chapter).

Because of its simplicity, radionuclide imaging to help determine whether a pericardial effusion is present is now frequently employed. The effusion must be of considerable size, however, before it can be demonstrated by radioactive methods. The detection of pericardial effusions through ultrasound has indeed been encouraging and may exceed the accuracy obtained with tracer techniques.

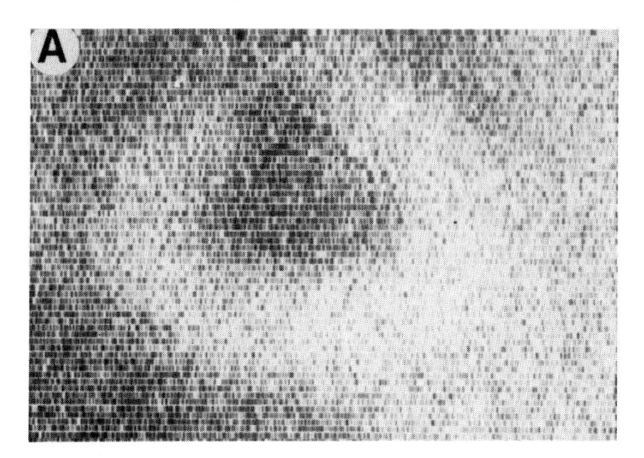

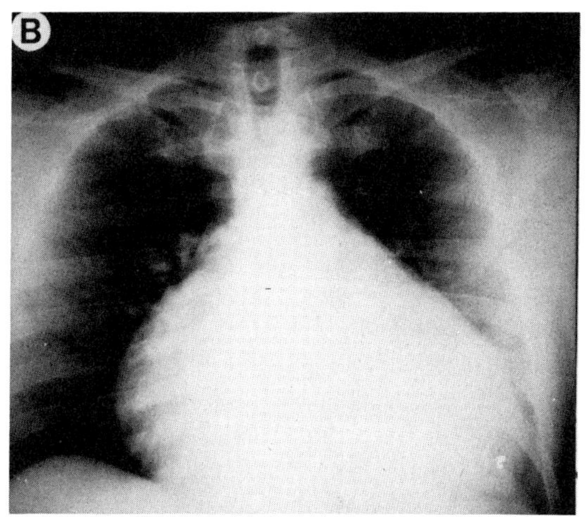

Fig. 15. Pericardial effusion. Note the marked disparity between the cardiac vascular pool on the scan (A) using ^{99m}Tc-HSA, and the cardiac silhouette on the x-ray (B), the latter outlined on the scan. (Courtesy of Dr. Ralph J. Gorten, University of Texas, Galveston.)

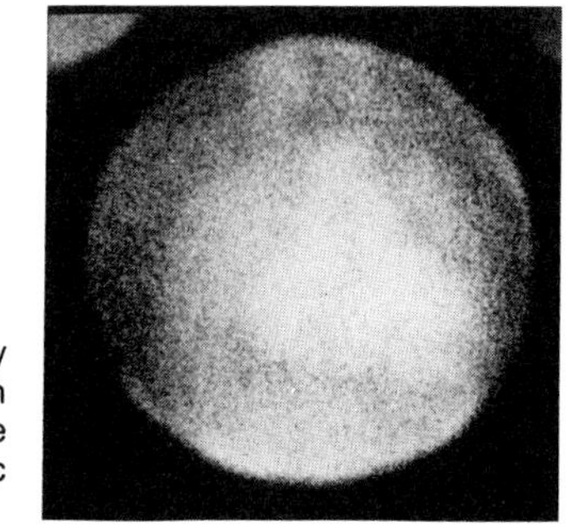

Fig. 16. Absence of pericardial effusion seen by scintillation camera after injection of ^{99m}Tc-sodium pertechnetate. Although the heart is enlarged, there is no appreciable separation between the cardiac and pulmonary images.

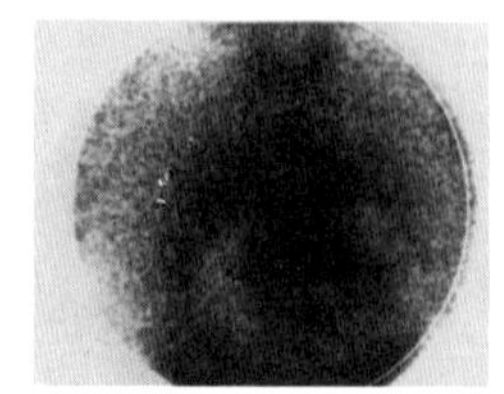

Fig. 17. Scintillation camera image of a patient with a pericardial effusion, using ^{99m}Tc-sodium pertechnetate. Note the wide separation between the cardiac vascular pool and pulmonary vasculature, indicating the site of the pericardial effusion.

IMAGING WITH ^{67}GA-GALLIUM CITRATE AND TUMOR-SEEKING AGENTS

From the very beginning of radionuclide imaging, an investigation has been underway for scanning agents that would be taken up preferentially in neoplasms. Although no ideal tumor-seeking agents have yet been found, the search for such radioactive materials continues to be a prime concern in nuclear medicine.

One of the most promising materials has been ^{67}Ga-gallium citrate. This radionuclide enters dividing cells in increased quantity when compared to the uptake in surrounding tissue.[10] Although it does show this specific, preferential uptake in tumor cells, it also accumulates in rapidly dividing cells that are not neoplastic; increased ^{67}Ga-concentration is seen, for example, in sarcoidosis, abscesses, pneumonia, and cirrhosis.

^{67}Ga has a half-life of 78 hours, and its gamma emissions at 184 keV (24 percent) and 296 keV (22 percent) may be utilized for scanning. Scanning is generally performed 72 hours after intravenous administration of 1.5 to 3.0 mCi ^{67}Ga-gallium citrate. The radioactive material leaves the blood slowly and accumulates in the liver, spleen, bone marrow, bone, and kidneys. The

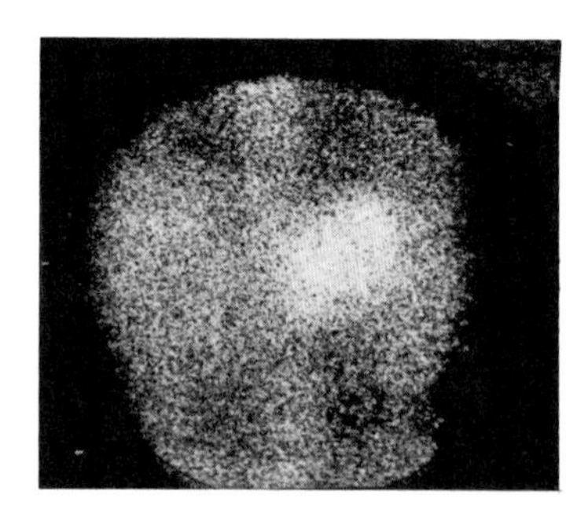

Fig. 18. ^{67}Ga-gallium citrate scan performed in the anterior position showing increased radioactivity to the left of the midline associated with abdominal involvement in Hodgkin's disease. (Courtesy of Dr. E.V. Furnas, Highland Hospital, Rochester, NY)

uptake in tumor occurs gradually, and tumor localization may be well demonstrated by 72 hours after injection.

Accumulation of ^{67}Ga in a neoplasm is seen in Figure 18; note the normal accumulation in the skeleton and liver. Imaging with radioactive gallium is frequently employed in staging the grades of malignant involvement, such as in Hodgkin's disease. Some other tumor-seeking agents are ^{111}In-indium chloride, ^{75}Se-selenomethionine, and ^{197}Hg-mercury chloride.

RADIOACTIVE INDIUM

A number of radioactive indium preparations have become available in recent years for nuclear medicine imaging.[11] The indium isotopes used clinically are ^{111}In and ^{113m}In.

^{113m}In has a relatively short half-life (99 minutes), is characterized by a monoenergetic gamma emission at 393 keV, and emits no beta radiation. It may be prepared in the laboratory by elution from a ^{113}Sn-^{113m}In generator system. The daughter ^{113m}In is present in the eluate in the trivalent form as ^{113m}In-chloride. Because the half-life of the parent ^{113m}Sn is relatively long (118 days), the generator can be used for 4 months before it needs to be replaced. Because of its relatively high energy gamma photon a rectilinear scanner should be utilized for ^{113m}In studies.

If carrier-free ionic radioactive indium (as chloride) at a pH 3 is injected intravenously, it becomes attached to plasma transferrin and apparently competes with iron for available binding sites in the transferrin. Circulating as a transferrin-^{113m}In complex, it finally enters the bone marrow and ultimately is incorporated into hemoglobin in a manner analogous to iron (Chapter 7).

The distribution and behavior of ^{113m}In-transferrin is similar to that of other labeled proteins. Thus it may be employed for such studies as radionuclide cisternography and cardiac vascular pool and placental imaging.[12] In regard to the latter, radioindium is confined for the most part to the maternal circulation with very little traversing the placenta to enter the fetal circulation.

^{113m}In-ferric hydroxide particles may be employed in perfusion lung scanning (Chapter 5). In addition, indium forms stable chelates with such substances as DTPA and has been used in this form for brain scanning.[13]

^{111}In has a relatively long half-life (64 hours). It is cyclotron-produced and emits photons in cascade at 173 keV and 247 keV. It is available in the trivalent form too as ^{111}In-chloride and becomes bound to transferrin when introduced intravenously at an acidic pH. Ionic ^{111}In is used principally to image the hematopoietic bone marrow and as a tumor-seeking agent. Since at least 24 hours should elapse before such imaging is undertaken, the long half-life of ^{111}In is more suitable than ^{113m}In for such studies.

Although a number of radiocolloids (e.g., ^{99m}Tc-sulfur colloid) are phagocytized in the bone marrow of the reticuloendothelial system (RES), transferrin-bound isotopes travel to the hematopoietic marrow rather than the reticuloendothelial marrow. Thus imaging of the marrow concerned with erythrocyte production may be accomplished with ^{111}In-indium chloride.[14] The marrow imaging may be performed 24 to 48 hours after injection of 1 to 2 mCi.

Although the mechanism is unknown, ^{111}In-transferrin has been found to

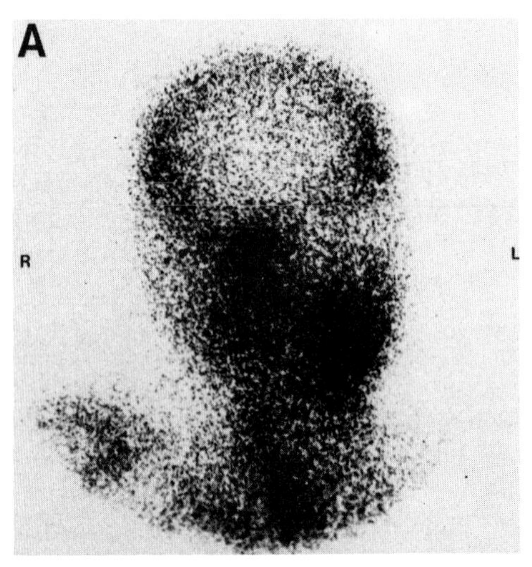

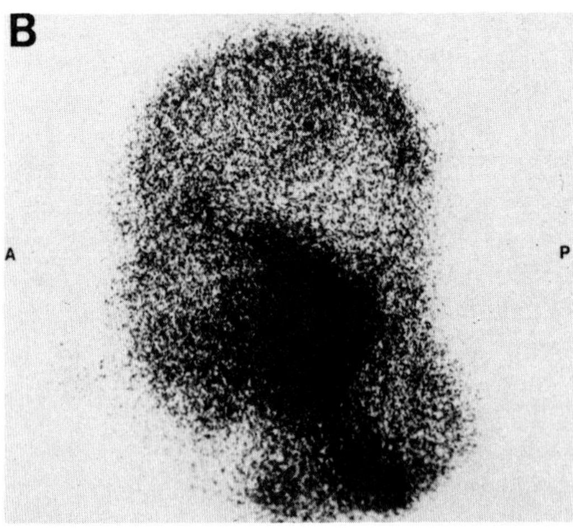

Fig. 19 Accumulation of ^{111}In-transferrin in a tumor (squamous cell carcinoma) involving the left submandibular region seen in the anterior (A) and left lateral (B) positions. (Courtesy of Dr. Paul A. Farrer, Royal Victoria Hospital, Montreal.)

localize in tumors.[15,16] Because the radionuclide enters the marrow, demonstration of ^{111}In in thoracic and pelvic tumors is not practical with this agent, but it is useful for imaging tumor involvement of the head and neck (Fig. 19).

Despite the knowledge that ionic radioindium becomes bound to transferrin and later is incorporated into erythrocytes, its ultimate fate is not well understood.[17]

^{75}SE-SELENOMETHIONINE STUDIES

Attempts to image the parathyroids and pancreas with radioactive materials have been disappointing. Although attractive in theory, endeavors to visualize these organs with ^{75}Se-selenomethionine have been far less successful than the results obtained, for example, when brain scanning with technetium 99m.

To prepare this radionuclide the sulfur atom in the methionine molecule is substituted with a gamma emitter, selenium 75. ^{75}Se-selenomethionine has many similarities to methionine, both chemically and biologically, and enters into protein synthesis.[18]

Following intravenous administration, ^{75}Se-selenomethionine is taken up rapidly by the parathyroid gland, pancreas, liver, and salivary glands. About 30 minutes after injection ^{75}Se-selenomethionine-bound plasma proteins begin to rise and may be detected with radioactive protein binding reaching a peak in 3 to 5 hours.

Behaving in a manner akin to methionine, the ^{75}Se-selenomethionine is incorporated into parathyroid polypeptides. In those cases of glandular hyperactivity, as in parathyroid hyperplasia or a functioning adenoma, the ^{75}Se-selenomethionine uptake is elevated and provides the basis for scan visualization.[19]

^{75}Se-selenomethionine is taken up in the thyroid gland too, and because of this it is necessary to administer a suppressive dose of triiodothyronine for 4 or 5 days prior to parathyroid imaging. The patient is given 250 μCi radioactive selenium intravenously, and the parathyroid scan is started within 10 minutes after injection. Because the radioactivity is very low four consecutive photoscans are obtained on a rectilinear scanner, and the scans are superimposed to enhance any localized areas of increased radioactivity that may be visualized. If the functioning adenoma or area of hyperplasia is of sufficient size (larger than 2 cm), visualization may be possible (Fig. 20). The number of false negatives is so high, however, that the reliability of this procedure is compromised.

Fig. 20 Parathyroid adenoma seen as a rounded area of increased radioactivity after ^{75}Se-selenomethionine injection by superimposition of four rectilinear scans. Arrows outline the adenoma on the identical image. The thyroid gland is faintly visualized.

Although pancreatic scanning has been enthusiastically received in a number of institutions, for the most part the results with this procedure are disappointing too.[20] The pancreas synthesizes protein at a relatively rapid rate in order to replenish the digestive enzymes elaborated by the gland. ^{75}Se-selenomethionine participates in this protein synthesis and is incorporated into the pancreas. Unfortunately the radionuclide is taken up in the liver too, further complicating scan interpretation.

Normal pancreatic scans are shown in Figures 21 and 22. No localized areas of absent radioactivity are noted. Using a subtraction technique the hepatic image may be deleted from the scan, with only the pancreas visualized making interpretation of the scan easier.

Carcinoma of the pancreas may be represented on the scan as an area of absent radioactivity (Fig. 23). Reduced uptake is also seen in pancreatitis. However, areas of absent radioactivity are so commonly encountered in patients with no pancreatic pathology that the value of the study is compromised. Since many tumors have an increased protein turnover, ^{75}Se-selenomethionine is sometimes employed as a tumor-seeking agent.

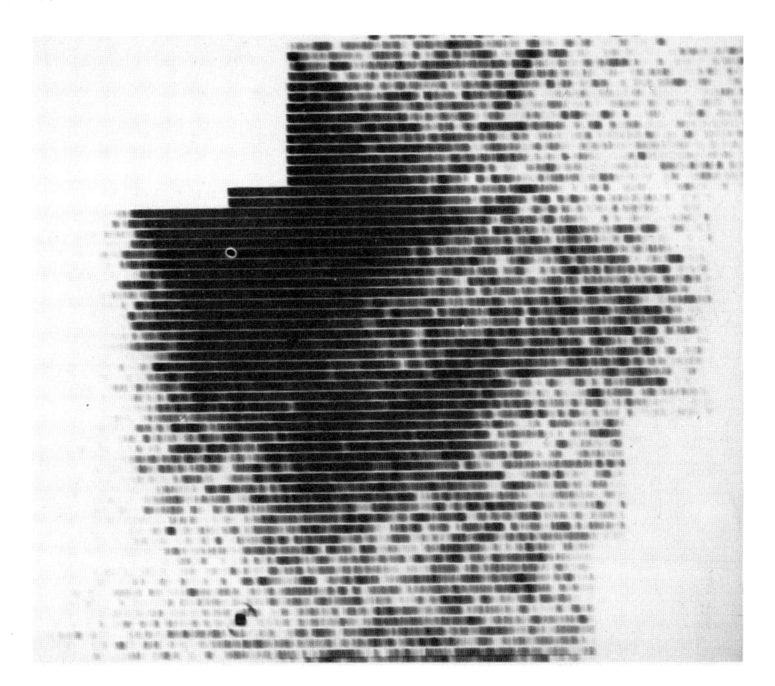

Fig. 21. Normal pancreatic scan after ^{75}Se-selenomethionine injection, performed on a rectilinear scanner. Note hepatic uptake of the radioactive material.

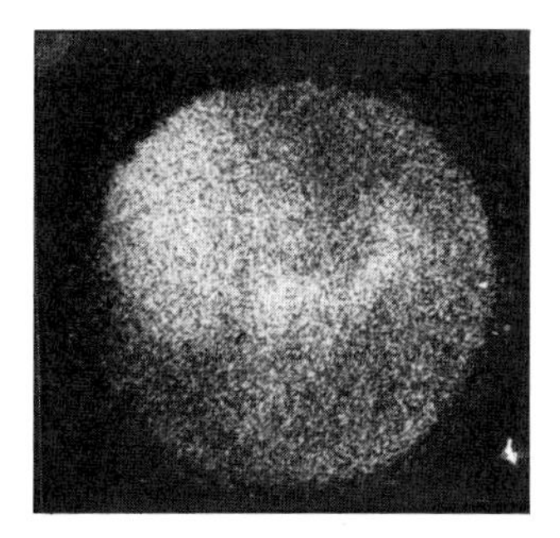

Fig. 22. Normal pancreatic imaging performed on a scintillation camera.

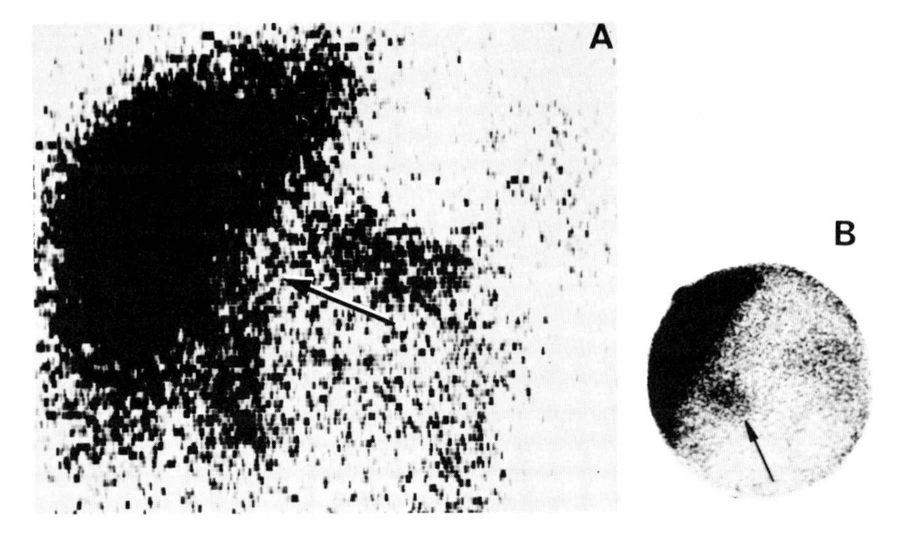

Fig. 23. Demonstration of carcinoma of the pancreas following ^{75}Se-selenomenthionine administration. A. Scan shows absent radioactivity (arrow) associated with carcinoma of the body of the pancreas. B. Imaging in a patient with neoplastic involvement of the body and tail of the pancreas, which shows evidence of radioactivity in only the head of the pancreas (arrow). (Courtesy of Dr. Said M. Zu'bi, Springfield Hospital Medical Center, Springfield, Mass.)

RADIONUCLIDE ANGIOGRAPHY

Practically unforeseen a decade ago, radionuclide angiography and venography have made amazing strides during the past several years. Radionuclide imaging of the course of a radioactive material as it traverses large vessels and perfuses certain organs is now performed regularly using a scintillation camera. Although studies with radioactive materials do not nearly approach the detail and structural information attained with roentgenologic methods, they do offer a relatively simple, noninvasive means by which patients may be screened, studied, and perhaps diagnosed without having to undergo the rigors that contrast angiography might entail.

At the present time radionuclide angiography is in a developmental phase. Theoretically, large vessel disease such as a major stenosis or aneurysm may be detected using radioactive materials. Most commonly, images are obtained at frequent intervals over the area in question following the injection of ^{99m}Tc-sodium pertechnetate.

A normal radionuclide angiocardiogram performed after injection of pertechnetate is shown in Figure 24. Exposures or frames were obtained at in-

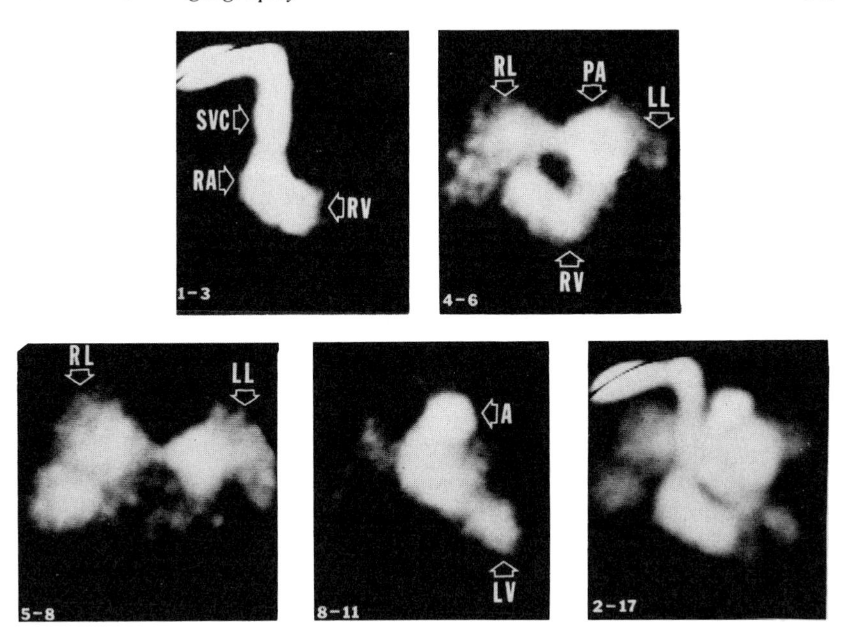

Fig. 24. Normal radionuclide angiocardiogram after ^{99m}Tc-sodium pertechnetate injection. The time intervals after radionuclide administration when imaging was obtained are noted in the lower left corner of each frame (first frame obtained during first to third seconds, second frame from fourth to sixth seconds, etc.). The last frame is a composite of images obtained between seconds 2 and 17. SVC, superior vena cava. RA, right atrium. RV, right ventricle. PA, pulmonary artery. LL, left lung. RL, right lung. LV, left ventricle. A, aorta. (Courtesy of Dr. Joseph P. Kriss, Stanford University Medical Center.)

tervals of 2 to 3 seconds. Following its injection into a peripheral vein in the right upper extremity, the structures through which the radioactive material courses are easily identified. In the first frame the superior vena cava, right atrium, and right ventricle are well seen. The pulmonary arterial bed is visualized in the third frame, and the left ventricle and aorta are clearly distinguished in the fourth. Unfortunately visualization of the aorta and its branches is less clear than that of the heart and entering veins. This is because much of the radioactive material has been diffusely perfused throughout the pulmonary vasculature, and by the time it leaves the aorta and its major branches the concentration of radioactivity is much reduced.

If a pericardial effusion is present, sequential imaging demonstrates a sizeable area of absent radioactivity between the heart and the pulmonary vascular bed (Fig. 25A). Static imaging in this case (Fig. 25B) shows the clear halo of absent radioactivity surrounding the cardiac vascular pool.

On the other hand, sequential imaging may assist in ruling out pericardial effusion in a patient whose chest x-ray suggests such a diagnosis.

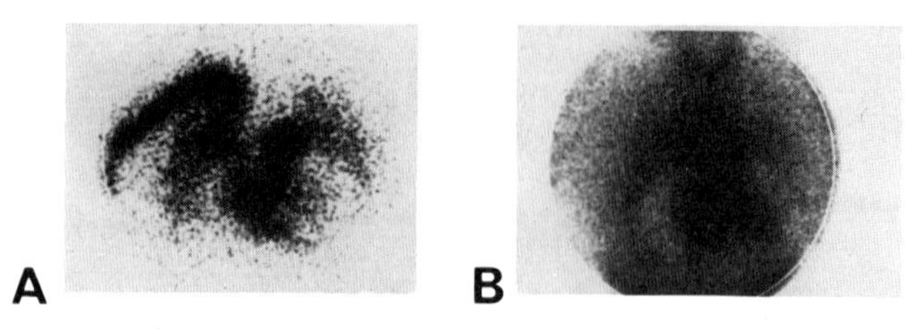

Fig. 25. Sequential imaging after ^{99m}Tc-sodium pertechnetate injection in a patient with a pericardial effusion. A. At 7.0 to 8.5 seconds. The radioactive material has now reached the lungs where a separation, due to a pericardial effusion, is seen between the cardiac blood pool and pulmonary vasculature. B. Static image demonstrates the pericardial effusion, seen as a halo of absent radioactivity surrounding the cardiac blood pool.

Radionuclide imaging of a patient with cardiac dilatation is illustrated in Figure 26. The image obtained during the interval 11.0 to 12.5 seconds (Fig. 26C) does not demonstrate a separation between cardiac and pulmonary vasculature. Failure to visualize this separation should effectively rule out pericardial effusion. Its absence is further confirmed by the static image seen in Figure 26D.

Several other items of interest may be demonstrated in this patient with cardiac dilatation. Since the radionuclide was injected in a left upper extremity vein, the course of ^{99m}Tc in the left subclavian and innominate veins and the superior vena cava is readily identified (Fig. 26A). Associated with the elevated venous pressure in this case, there is reflux of radionuclide into the left internal jugular vein (Fig. 26B).

Imaging of patients with superior venal caval obstruction due to neoplasm may demonstrate multiple venous collaterals (Fig. 27). Radionuclide imaging to demonstrate cardiac defects in both congenital and acquired heart disease is now carried out in a number of medical centers. Largely through the efforts of several outstanding investigators there have been encouraging advances in this area.[21] Abnormalities associated with congenital cardiac disorders have been imaged with a remarkable degree of anatomic detail. Radionuclide angiography in a patient with a left ventricular aneurysm, before and after corrective surgery, is shown in Figure 28.

Sequential imaging has been useful in determining if a mass is solid (e.g., tumor) or whether it is associated with an aneurysmal dilatation of a large artery. Consider a patient found to have a mass on chest x-ray (Fig. 29A) and whose radionuclide angiogram is presented in Figure 29B. Sequential imaging 11.0 to 12.5 seconds after ^{99m}Tc injection shows no evidence of increased perfusion at the site of the abnormality. These findings suggest the absence of a lesion, such as a functioning aneurysm; at surgery bronchogenic carcinoma was found. On the other hand, the abnormality on the x-ray shown in Figure

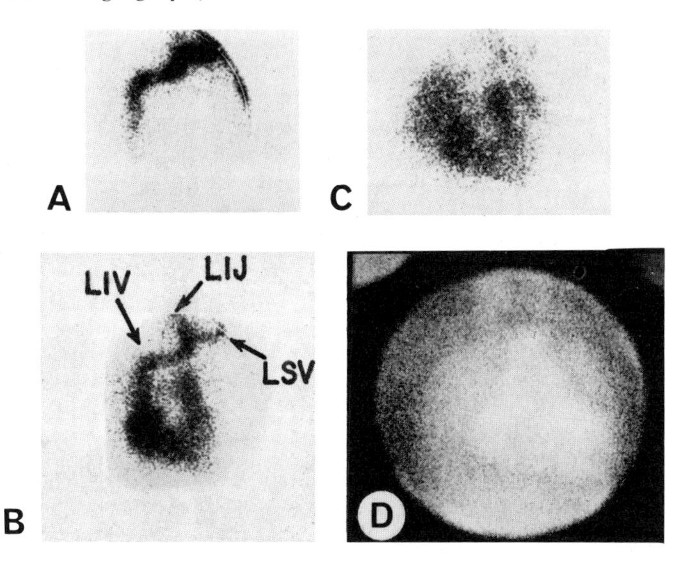

Fig. 26. Sequential imaging after pertechnetate injection in a patient with cardiac dilatation. After radionuclide injection in a left upper extremity vein, frames were obtained at intervals of 1.5 seconds. A. At 2.0 to 3.5 seconds. Radionuclide is seen in the left subclavian and innominate veins entering the superior vena cava. B. At 6.5 to 8.0 seconds. The right atrium and right ventricle are well visualized, and the radionuclide is now seen in the pulmonary artery. Note the reflux of ^{99m}Tc into the left internal jugular vein. C. At 11.0 to 12.5 seconds. As the radioactive material appears in the lungs, observe the absence of separation between the pulmonary and cardiac radioactivity. D. Static image. There is no evidence of a pericardial effusion. The pulmonary and hepatic blood pool images are closely related to the cardiac image with no discernible separation. LIJ, left internal jugular vein. LSV, left subclavian vein. LIV, left innominate vein.

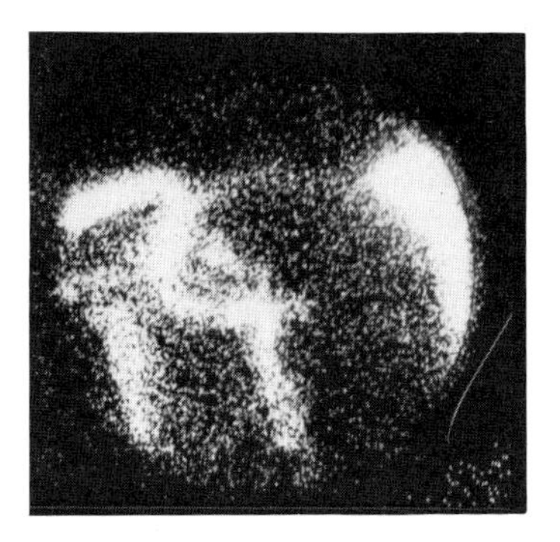

Fig. 27. Rapid sequential imaging in a patient with superior vena cava obstruction due to tumor. Taken during the interval 5 to 7 seconds. Numerous venous pathways involving internal mammary and intercostal veins are demonstrated. The radionuclide was injected simultaneously in both upper extremities.

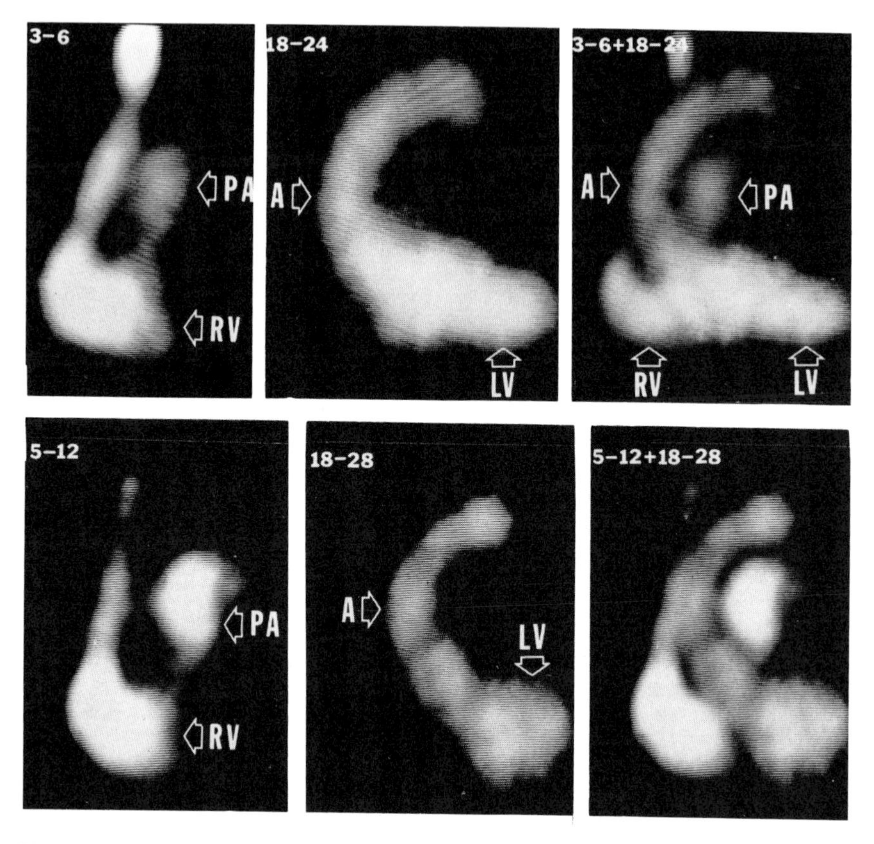

Fig. 28. Left ventricular aneurysm before (top three frames) and after (bottom frames) surgical correction, following administration of ^{99m}Tc-sodium pertechnetate. The time interval after ^{99m}Tc injection during which imaging was obtained is noted in the upper left corner of each frame. The aneurysm of the left ventricle is well seen in interval (second frame) and composite (third frame) preoperative images (at top). Postoperative imaging shows absence of the aneurysm after surgery. (Courtesy of Dr. Joseph P. Kriss, Stanford University Medical Center.)

30A is indeed an aneurysmal dilatation. This is demonstrated on the image in Figure 30B.

Visualization of arterial structures with radionuclide studies results in images with far less clarity than the large veins. However, it is frequently possible to demonstrate successfully major blood vessel occlusion such as that seen with partial senosis of the right internal carotid artery (Fig. 31).

Radionuclide angiography of the internal carotid artery and its branches may be performed in conjunction with brain imaging. Occlusion of the internal carotid artery may be visualized directly or associated with absent perfu-

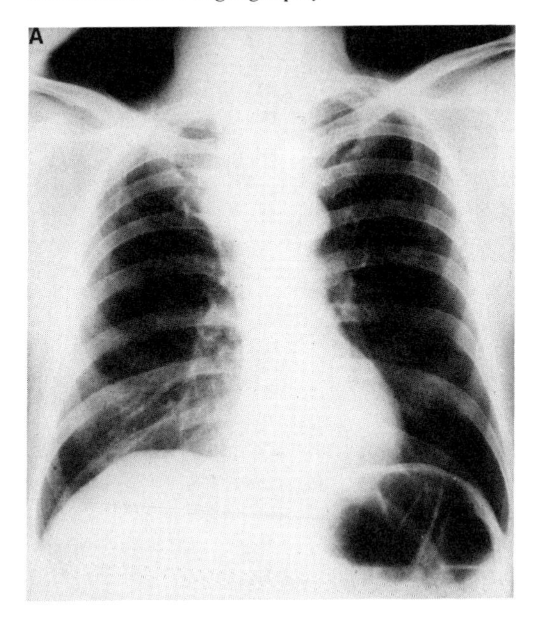

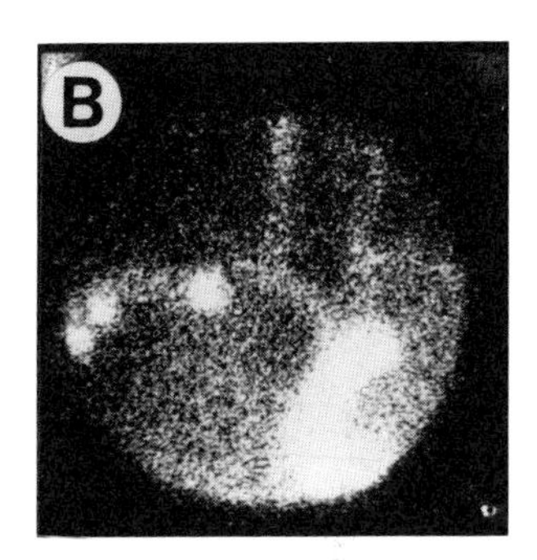

Fig. 29. Absence of aneurysm. A. Chest x-ray showing right-sided density. B. Radionuclide angiogram during interval of 11.0 to 12.5 seconds after ^{99m}Tc injection. The aortic arch and its branches are easily identified. Failure to demonstrate a vascular lesion at the site of the abnormality seen on the x-ray indicates the absence of a functioning aneurysm. At surgery the lesion was found to be a bronchogenic carcinoma.

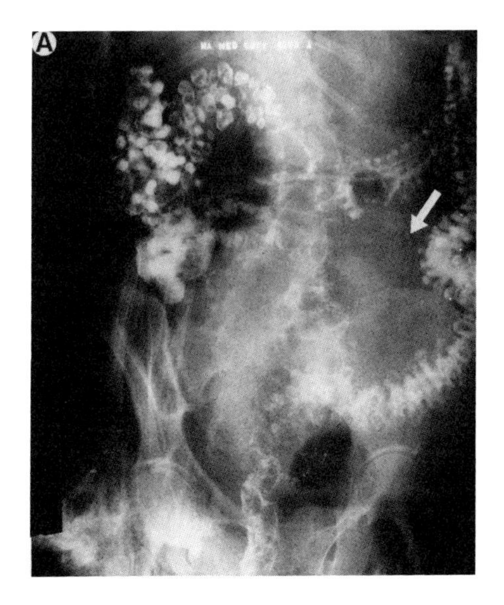

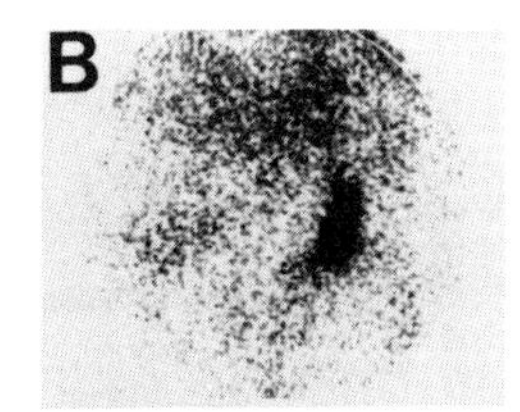

Fig. 30. Presence of aneurysm. A. Abdominal x-ray showing evidence of aortic aneurysm (arrow). B. Increased radioactivity associated with the aneurysm after ^{99m}Tc-sodium pertechnetate injection.

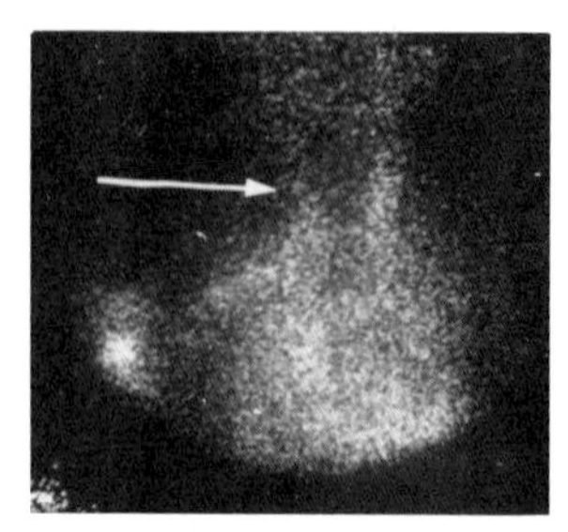

Fig. 31. Partial occulsion of the right internal carotid artery (arrow) demonstrated during the interval of 12.0 to 13.5 seconds after ^{99m}Tc administration.

sion of the areas supplied by the anterior and middle cerebral arteries (Fig. 32).

The limitations of radionuclide angiography as now employed were mentioned earlier in this section. Certainly corrective heart surgery or resection of an aortic aneurysm must not be undertaken on the basis of radionuclide studies alone. However, tracer angiography is extremely helpful when direction is needed to help determine which of the more sophisticated diagnostic procedures should be undertaken. If, for example, it is difficult to

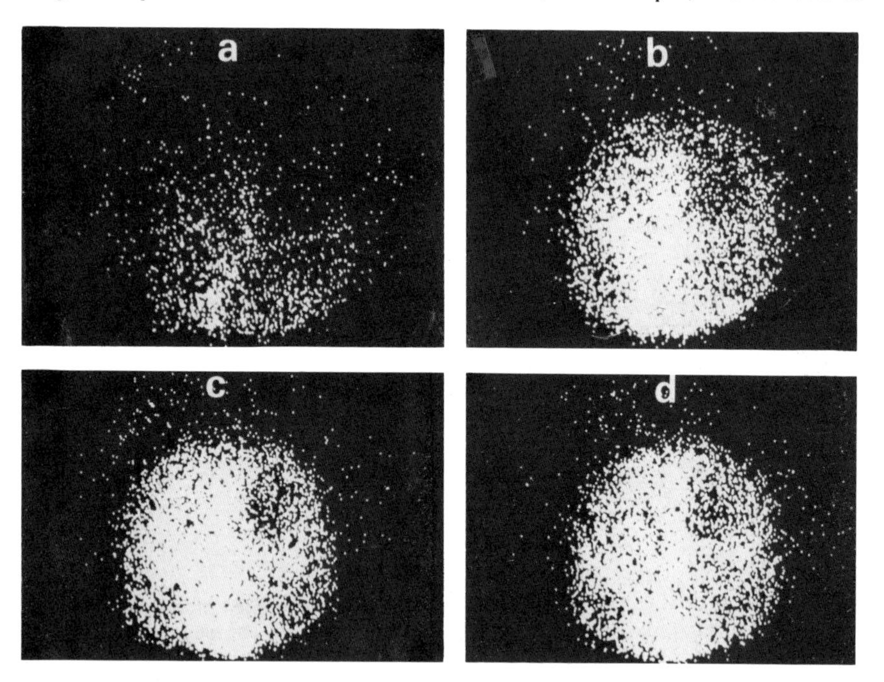

Fig. 32. A. Sequential brain imaging in the anterior position in a patient with carotid artery stenosis and relatively absent initial perfusion in the distribution of the left anterior and middle cerebral arteries. A. Interval 8 to 10 seconds. There is increased perfusion to the right side, relative absence on the left.

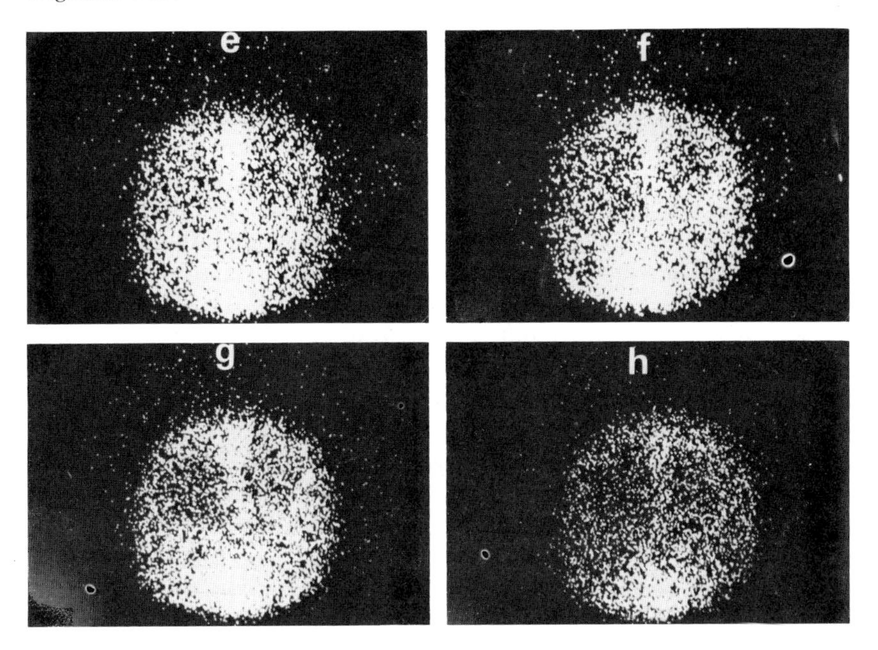

Fig. 32. B. Interval 18 to 20 seconds. Crossover from the right side shows further left-sided perfusion in distribution of left middle cerebral artery. (Courtesy of Dr. Paul A. Farrer, Royal Victoria Hospital, Montreal.)

ascertain whether a mass on a chest x-ray is tumor or an aneurysmal dilatation, a screening radionuclide angiogram helps determine the general nature of the abnormality. This in turn points toward the next appropriate procedure (for a tumor perhaps tomography, for an aneurysm perhaps contrast arteriography). Screening for such entities as carotid artery stenosis is frequently employed.

In many instances radionuclide angiography may be used to help establish a clinical diagnosis when more elaborate studies or radical surgery are neither contemplated nor desired. Thus the presence of an asymptomatic large aneurysm of the descending aorta may simply be confirmed with radioactive methods.

REGIONAL BLOOD FLOW

The use of radioactive materials in evaluating regional blood flow has been applied to a number of organs and systems. However, it should probably still be considered in the area of clinical investigation rather than an established nuclear medicine procedure. It is beyond the scope of this book to explore this subject in detail, as it involves the use of complex mathematical analyses. The interested reader should consult the references at the end of this chapter.

A wide variety of radioactive materials have been used in blood flow determinations. Particularly attractive are the radionuclides that are readily diffusible in fat, other body tissues, and capillary walls. This group is best exemplified by xenon 133. This radioactive gas is easily dissolved in saline and may be administered parenterally in this form.

In most procedures, estimating blood flow involves the relationship between the radioactivity delivered to an organ or region and the radioactivity that remains over a given period of time. The radionuclide may be injected into the structure itself under study (e.g., skeletal muscle or uterus) or it may be given directly into a blood vessel (e.g., internal carotid artery when estimating cerebral blood flow).

Suppose one wishes to determine the blood flow in a lower extremity muscle. If a small amount of ^{133}Xe dissolved in saline (500 μCi in 0.5 ml) is injected into a leg muscle, continual monitoring over the site of injection provides information on how rapidly the radioactive material is leaving the area.[22] The extent of reduction in radioactivity over a given period of time is related to the rate at which the highly soluble ^{133}Xe leaves the area as it enters the capillaries and is ultimately excreted in the lungs. The more rapid the reduction in radioactivity, the greater is the blood flow from the extremity; a slower fall in radioactivity is indicative of reduced blood flow. The blood flow per minute per unit weight can be calculated from an analysis of the decline in radioactivity and an estimate of muscle mass. This method has been used to determine the reduction of muscle blood flow in such entities as diabetes and arteriosclerosis obliterans.

For cerebral blood flow studies the ^{133}Xe solution may be injected directly into the internal carotid artery. Since the diffusible tracer leaves the brain capillaries rapidly and almost entirely to enter the tissue spaces, the rate of its removal in venous blood, as determined by external monitoring over the specific area in question, may be calculated.[23]

In a related area, relatively large caliber particles tagged with a tracer have been used to determine the distribution of regional blood flow and to help detect the presence of shunts or anastomoses.[24] Particles are prepared whose diameters are larger than those of the capillaries of the organ to be studied. After these particles are injected into a major blood vessel feeding the organ they travel as far as the capillaries, where they are filtered out. If there are arteriovenous shunts of sufficient diameter to allow the large particles to pass through them, the degree to which such shunting occurs is then determined.

Such methods in humans are most commonly applied in perfusion lung scanning; here large particles (e.g., ^{99m}Tc-albumin microspheres, diameter 30 μ) are injected into a peripheral vein and are filtered out in the pulmonary capillary bed in accordance with the pulmonary arterial distribution. Related procedures have wide application in the experimental laboratory, such as using tagged glass spherules to help identify collateral vessels after coronary artery occlusion in dogs.

REFERENCES

Bone Imaging

1. Bachman AL, Sproul EE: Correlation of radiographic and autopsy findings in suspected metastases in the spine. Bull NY Acad Med 31:146, 1955
2. Spencer R, Herbert R, Rish MW, et al: Bone scanning with ^{85}Sr, ^{87m}Sr and ^{18}F. Br J Radiol 40:641, 1967
3. Charkes ND: Bone scanning: principles, techniques and interpretation. Radiol Clin North Am 8:259, 1970
4. Harmer CL, Burns JE, Sams A, et al: The value of fluorine-18 for scanning bone tumours. Clin Radiol 20:204, 1969
5. Subramanian G, McAfee JG, Bell EG, et al: ^{99m}Tc labeled polyphosphate as a skeletal imaging agent. Radiology 102:701, 1972

5.a. Castronovo FP Jr, Callahan RJ: New bone scanning agent: ^{99m}Tc-labeled 1-hydroxyethyliden-1, 1-disodium phosphanate. J Nucl Med 13:823, 1972

5.b. Pendergrass HP, Potsaid MS, Castronovo FP Jr: The Clinical use of ^{99m}Tc-diphosphonate (HEDSPA). Radiology 107:557, 1973

Placentography

6. Cooper RD, Izenstark JL, Branyon DL Jr, et al: Isotope placentography or placental imaging? Radiology 87:291, 1966
7. Nelp WB, Larson SM: Diagnosis of placenta praevia by photoscanning with albumin labeled with technetium Tc 99m. JAMA 200:148, 1967

Imaging for Pericardial Effusions

8. Bonte FJ, Curry TS III: The cardiovascular blood pool scan. In Freeman LM, Johnson PM (eds): Clinical Scintillation Scanning, New York, Harper & Row, 1969, pp 203-221
9. Staab EV, Patton DD: Nuclear medicine studies in patients with pericardial effusion. Semin Nucl Med 3:177, 1973

^{67}Ga-Gallium Citrate and Tumor-Seeking Agents

10. Edwards CL, Hayes RL: Scanning malignant neoplasms with gallium 67. JAMA 212:1182, 1970

Radioactive Indium

11. Stern HS, Goodwin DA, Scheffel U, et al: In^{113m} for blood-pool and brain scanning. Nucleonics 25:62, 1967
12. Niehoff RD, Hendee WR, Brown DW: Placenta scanning with ^{113m}In. J Nucl Med 11:15, 1970
13. O'Mara RE, Subramanian G, McAfee JG, et al: Comparison of ^{113m}In and other short-lived agents for cerebral scanning. J Nucl Med 10:18, 1969
14. Lilien DL, Berger HG, Anderson DP, et al: ^{111}In-chloride: a new agent for bone marrow scanning. J Nucl Med 14:184, 1973
15. Goodwin DA, Goode R, Brown L, et al: ^{111}In-labeled transferrin for the detection of tumors. Radiology 100:175, 1971
16. Farrer PA, Saha GB, Shibita HN: Evaluation of ^{111}In-transferrin as a tumor scanning agent, J Nucl Med 13:429, 1972
17. Castronovo FP Jr, Wagner HN Jr: Comparative toxicity and pharmacodynamics of ionic indium chloride and hydrated indium oxide. J Nucl Med 14:677, 1973

^{75}Se-Selenomethionine Studies

18. Blau M, Holland JF: Metabolism of selenium-75 l-selenomethionine. In Andrews GA, Kniseley RM, Wagner HN Jr (eds): Radioactive Pharmaceuticals. Oak Ridge, Tenn, US Atomic Energy Commission, 1966, pp 423-428
19. Potchen EJ, Watts HG, Awwad HK: Parathyroid scintiscanning. Radiol Clin North Am 5:267, 1967
20. Landman S, Polcyn RE, Gottschalk A: Pancreas imaging—is it worth it? Radiology 100:631, 1971

Radionuclide Angiography

21. Kriss JP, Enright LP, Hayden WG, et al: Radioisotope angiocardiography: findings in congenital heart disease. J Nucl Med 13:31, 1973

Regional Blood Flow

22. Lassen NA, Lindbjerg IF, Munck O: Measurement of blood flow through skeletal muscle by intramuscular injection of ^{133}xenon. Lancet 1:686, 1964
23. Oldendorf WH: Measurement of cerebral blood flow. In Blahd WH (ed): Nuclear Medicine, 2nd ed. New York, McGraw-Hill, 1971, pp 294-301
24. Wagner HN Jr, Rhodes BA, Sassaki Y, et al: Studies of the circulation with radioactive microspheres. Invest Radiol 4:374, 1969

PART TWO

The Nuclear Medicine Laboratory

chapter 7

SPECIAL LABORATORY PROCEDURES

BLOOD VOLUME

For the most part, the measurement of bodily compartments with radioactive materials is performed by determining the extent to which a tracer becomes diluted in the particular compartment under study. Measuring the intravascular space, or circulating blood volume, is by far the most commonly performed test in this group of procedures.

The volume of fluid in a closed compartment or container can be determined with the use of radioactive materials if a tracer, upon being added to the fluid, is distributed uniformly. If the total radioactivity of the added radionuclide and the radioactivity per milliliter of the fluid after the tracer has become evenly dispersed are known, the volume in the container can be readily calculated.

Consider a liquid of unknown volume in a container as shown in Figure 1; 2 ml ^{131}I-HSA (5 $\mu Ci/ml$) was added to the fluid and the container agitated lightly to ensure uniform distribution of the ^{131}I-HSA. After mixing was completed 15 minutes later, a sample of fluid was obtained. The radioactivity was determined for both the tracer that was added and for 1 ml of the liquid containing the added tracer. By determining the extent to which the tracer became diluted when mixed evenly in the fluid, the volume of the liquid in the container can be calculated.

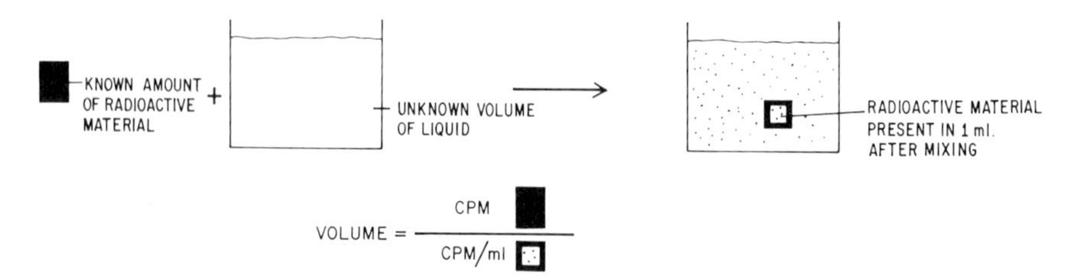

Fig. 1. Container holding liquid of unknown volume to which was added a known amount of radioactive material (dots) uniformly distributed in the fluid. The volume of the liquid is calculated by determining the extent to which the tracer is dispersed.

$$\text{Volume} = \frac{\text{radioactivity of tracer}}{\text{radioactivity/ml of fluid after uniform distribution of tracer}}$$

$$\text{Volume} = \frac{2 \text{ ml} \times 10{,}000 \text{ cpm/ml}}{40 \text{ cpm/ml}} = 500 \text{ ml}$$

Similar principles may be applied clinically. When determining the blood volume with radioactive materials, a radionuclide is employed that after intravenous administration should ideally be (1) confined to the circulatory system and (2) uniformly distributed in the intravascular space in a relatively short period of time (15 to 30 minutes). Unfortunately the intravascular space is not a closed compartment and small portions of some of the materials used for blood volume studies may escape from it.

It is essential that the precise volume of the solution containing the tracer be known. A small error in determining the volume of the material injected is greatly magnified when the blood volume is ultimately calculated.

As with all other procedures that measure radioactivity in blood samples, it should be ascertained if the patient has undergone other radionuclide examinations in the recent past. If it is affirmative, a blood sample must be obtained prior to the blood volume study to determine the initial blood radioactivity so it may be subtracted from the values obtained after tracer administration.

The terms *corrected hematocrit* and *corrected plasmacrit* are commonly encountered when dealing with blood volume determinations. The concentration of erythrocytes in the peripheral venous blood is appreciably higher than in the arterial system and in several organs (e.g., the liver). The term *total body hematocrit* refers to the average hematocrit present in the human which is approximately 92 percent of the venous hematocrit. The result obtained is also known as the corrected hematocrit.

When centrifuging a sample of whole blood to separate the plasma and erythrocytes, a portion of the plasma becomes enmeshed in the cells. This varies from 1 percent when using small bore capillary tubes to 4 percent when the larger Wintrobe tubes are employed. To adjust for this phenomenon the plasmacrit (1-Hct) is multiplied by 99 or 96 percent, depending on the type of tube used, to obtain the corrected plasmacrit.

It is common practice now to perform capillary or microhematocrits rather than using Wintrobe tubes. Because multiplying the observed plasmacrit by 99 percent would not significantly alter its value, this additional calculation has not been included in the sections dealing with blood volume calculations.

A number of radioactive materials have been successfully employed in studying the blood volume in humans. Those most frequently used are radioiodinated human serum albumin (^{125}I or ^{131}I-HSA) and ^{51}Cr-labeled erythrocytes.

^{131}I-HSA Studies

One of the problems in using ^{131}I-HSA for blood volume determinations is that it does not remain entirely confined to the circulatory system.[1] If a small quantity of this radioactive material is injected intravenously it undergoes uniform distribution for the most part in a manner similar to other proteins present in plasma, but a small amount of the tagged albumin "leaks" to the lymph and interstitial fluid. By the time there is uniform mixing of the ^{131}I-HSA in blood, some 15 to 20 minutes after injection, approximately 2 percent of the tracer has escaped from the intravascular space. However, if the small error associated with this loss of tagged material from the circulation can be ignored, ^{131}I-HSA may be employed for both whole blood and plasma volume determinations.

The dose of ^{131}I-HSA used in these studies may be relatively small, usually 5 to 10 μCi. It is common practice to prepare a solution of ^{131}I-HSA in a concentration of 5 μCi/ml.

Whole Blood Volume. The ^{131}I-HSA is injected into a peripheral vein; 15 minutes later, after mixing is complete, a blood sample is taken from the opposite arm and placed in an oxalated tube. The blood sample should be inverted from time to time to ensure uniform distribution of the tagged albumin in the blood. The radioactivity is determined for both the injected ^{131}I-HSA and for 1 ml of the withdrawn blood, and the blood volume calculated from this information.

Because the radioactivity per milliliter of the tracer is usually quite high, counting errors may be avoided by measuring a diluted sample of the tracer: 1 ml of the tracer is placed in 500 or 1000 ml of water and the radioactivity per milliliter of the diluted ^{131}I-HSA determined. Using the specific dilution factor employed, the tracer's radioactivity can then be determined.

Example. The patient was given an injection of 2 ml ^{131}I-HSA solution containing 5 μCi/ml, and an oxalated blood sample was obtained 15 minutes later.

$$\text{Radioactivity/ml of tracer sample (1 ml diluted to 1000 ml)} = 120 \text{ cpm/ml}$$

$$\text{Radioactivity/ml whole blood} = 40 \text{ cpm/ml}$$

$$\text{Blood volume} = \frac{\text{radioactivity of injected tracer}}{\text{radioactivity/ml of blood after mixing}}$$

$$= \frac{2 \text{ ml} \times 120 \text{ cpm/ml} \times 1000}{40 \text{ cpm/ml}} = 6000 \text{ ml}$$

The 1 ml of the initial tracer was diluted to 1000 ml before counting, so the radioactivity per milliliter of the tracer is calculated by counting 1 ml of the diluted sample and multiplying it by the dilution factor (1000). Bcause 2 ml was injected into the patient, the radioactivity per milliliter tracer is multiplied by 2.

It is possible to calculate the plasma volume after determining the whole blood volume, but this is not as accurate as the direct determination of plasma volume itself. This calculation can be made by multiplying the whole blood volume (obtained with ^{131}I-HSA) by the corrected venous plasmacrit.

Plasma Volume. Because ^{131}I-HSA behaves in a manner akin to the plasma proteins, it is convenient to use this material to measure the plasma volume. The procedure followed is similar to that described for the ^{131}I-HSA blood volume determination except that the 15-minute blood sample is centrifuged so the radioactivity in the plasma may be determined.

Example. The patient was given an injection of 2 ml ^{131}I-HSA solution containing 5 μCi/ml, and an oxalated blood sample was obtained 15 minutes later. The blood specimen was centrifuged and the plasma was pipetted off.

$$\text{Radioactivity/ml of tracer sample (1 ml diluted to 1000 ml)} = 120 \text{ cpm/ml}$$

$$\text{Radioactivity/ml plasma} = 15 \text{ cpm/ml}$$

$$\text{Plasma volume} = \frac{\text{radioactivity of injected tracer}}{\text{radioactivity/ml plasma after mixing}}$$

$$= \frac{2\text{ ml} \times 120\text{ cpm/ml} \times 1000}{70\text{ cpm/ml}} = 3429\text{ ml or } 3.4\text{ liters}$$

^{51}Cr-Labeled Erythrocyte Studies

The use of tagged erythrocytes provides an excellent method for studying the blood volume with a substance that is confined to the intravascular space.[2] Since the chromate ion penetrates the erythrocyte cell membrane and becomes firmly bound to hemoglobin, erythrocytes may be easily labeled with ^{51}Cr-sodium chromate.

A sample of the patient's blood is withdrawn (10 ml) and placed in a vial containing an anticoagulant solution, acid-citrate-dextrose (ACD). To this is added approximately 25 μCi ^{51}Cr-sodium chromate, and the mixture is allowed to stand at room temperature, with occasional gentle agitation, for 30 minutes as the labeled chromate tags the erythrocytes. ACD solution is used because the entry of chromate into the cells occurs more quickly and to a greater extent in an acidic medium. By 30 minutes well over 90 percent of the erythrocytes are radioactively labeled.

Chromate ($CrO_4^{=}$ or $Cr_2O_7^{=}$) freely penetrates the cell membrane and is bound to hemoglobin when the valence of ^{51}Cr is + 6, but such labeling does not occur when the ^{51}Cr valence is +3. In the latter form the chromate may become bound to circulating hemoglobin or plasma proteins but does not tag the erythrocytes.[3]

Therefore after the 30-minute incubation 1 ml ascorbic acid, a reducing agent, is added and the mixture is allowed to stand at room temperature another 15 minutes. The purpose of this step is to reduce the valence of ^{51}Cr from +6 to +3 to take care of the small quantity of ^{51}Cr-sodium chromate that is still free so it does not become available to bind circulating erythrocytes when the tagged mixture is injected.

The labeled erythrocyte solution may be referred to as the standard. Putting aside 6 ml for laboratory determinations, the remainder of the standard, with volume accurately determined, is injected into the patient intravenously with a heparinized syringe. After 15 to 30 minutes a 10-ml blood sample is obtained from a vein in the opposite arm. If the patient has marked splenomegaly or hepatomegaly, 30 to 60 minutes should be allowed for mixing.

The initial objectives are to determine (1) the radioactivity due to the erythrocytes injected into the patient and (2) the extent to which these labeled cells become diluted in the circulation after uniform mixing has occurred some 15 to 30 minutes later by finding out the radioactivity present in 1 ml of the cells. First the erythrocyte mass is calculated, and then from this the blood volume is determined.

Radioactive counts are obtained on the standard, supernatant of the standard after centrifugation, and the whole blood and plasma of the withdrawn blood sample. The hematocrit of both the standard and blood samples are also determined.

The following symbols may be used:

V = volume of standard injected (ml)
Std = cpm/ml standard
Std_{sup} = cpm/ml supernatant of standard
Pt = cpm/ml whole blood of patient
Pt_{pl} = cpm/ml plasma of patient
Hct_{std} = hematocrit of standard
$1\text{-}Hct_{std}$ = plasmacrit of standard
Hct_{pt} = hematocrit of patient
$1\text{-}Hct_{pt}$ = plasmacrit of patient

The erythrocyte (RBC) mass may be obtained from this formula:

$$\text{RBC mass} = \frac{\text{total radioactivity of tagged RBCs injected}}{\text{radioactivity/ml of RBCs after mixing}}$$

To determine the radioactivity of the injected ^{51}Cr-labeled RBCs, it is necessary to find out the radioactivity in the standard that is due to the RBCs themselves. If the radioactivity in 1 ml supernatant of the standard (Std_{sup}) is known, multiplying this figure by the plasmacrit of the standard ($1\text{-}Hct_{std}$) gives the radioactivity present in 1 ml whole standard that is due to the supernatant ($Std_{sup} \times 1\text{-}Hct_{std}$). If this is subtracted from the radioactivity present in 1 ml whole standard (Std), the radioactivity due to the RBCs present in 1 ml standard can be found ($[Std_{sup} \times 1\text{-}Hct_{std}]$). Then by multiplying the latter expression by the volume of the standard (in milliliters) that was injected (V), the total radioactivity of tagged RBCs injected is obtained.

In a similar fashion, the radioactivity in the patient's blood after the tagged RBCs are uniformly distributed may be calculated. The radioactivity in 1 ml plasma (Pt_{pl}) is found, and this is multiplied by the plasmacrit of the patient ($1\text{-}Hct_{pt}$) to find the radioactivity due to the plasma present in 1 ml whole blood. This value ($Pt_{pl} \times 1\text{-}Hct_{pt}$) is subtracted from the radioactivity in 1 ml whole blood; the result is the radioactivity due to the erythrocytes present in 1 ml whole blood ($Pt - [Pt_{pl} \times 1\text{-}Hct_{pt}]$). To determine the radioactivity that would be present in 1 ml erythrocytes this expression is divided by the patient's hematocrit.

Using the aforementioned terms, the formula for the RBC mass may now be given as:

$$\text{RBC mass} = \frac{V \times [\text{Std} - (\text{Std}_{sup} \times 1 - \text{Hct}_{std})]}{\dfrac{\text{Pt} - (\text{Pt}_{pl} \times 1 - \text{Hct}_{pt})}{\text{Hct}_{pt}}}$$

$$\text{RBC mass} = \frac{V \times [\text{Std} - (\text{Std}_{sup} \times 1 - \text{Hct}_{std})] \times \text{Hct}_{pt}}{\text{Pt} - (\text{Pt}_{pl} \times 1 - \text{Hct}_{pt})}$$

The whole blood volume may then be determined by dividing the RBC mass by the total body hematocrit $(0.92 \times \text{Hct}_{pt})$.

As an alternative to the method described, the standard may be centrifuged and the erythrocytes washed and resuspended in normal saline prior to injection to get rid of unbound ^{51}Cr, ACD solution, and ascorbic acid. However, the possible damage to the cells may outweigh any advantage gained by this procedure.

Clinical Considerations. Studies with ^{51}Cr-labeled erythrocytes have both a firm theoretical basis and, when properly performed, a high degree of accuracy. Although more difficult to perform, there is no question that this is the procedure of choice if a reliable measurement of the circulating blood volume or erythrocyte mass is desired. This is especially true in those cases in which it is essential that the determination of this mass is accurate, as in patients with polycythemia vera.

Normally the erythrocyte volume is approximately 29 ml/kg in men and 25 ml/kg in women.

ERYTHROCYTE (RED BLOOD CELL) SURVIVAL

A number of procedures have been devised for the determination of RBC survival time using cells labeled with radioactive materials.[4] The RBC life span may be inferred from the information obtained in these studies. Tagging the cells is accomplished by one of two methods: random labeling of RBCs or labeling a cohort of cells. The former is the procedure most commonly employed clinically.

Random Labeling

The random labeling of RBCs involves obtaining a sample of blood from a patient and tagging the cells in the specimen with a radionuclide. The RBCs are of different ages and are labeled at random, as the name of the procedure implies. The radioactive materials most commonly used for this purpose are ^{51}Cr-sodium chromate and ^{3}H or ^{32}P-diisopropylfluorophosphate (DFP); most RBC survival studies are done with ^{51}Cr.

^{51}Cr-Sodium Chromate Studies. The cells are labeled with ^{51}Cr in a manner identical to that described for ^{51}Cr blood volume studies, except that a larger dose of radioactive material is used, usually 250 to 300 μCi. The entire radioactive mixture is injected intravenously, and the RBC survival study is accomplished through (1) analysis of the radioactivity in blood samples and (2) external monitoring over the spleen, liver, and heart.

Blood Studies. There are several important characteristics related to the binding of radiochromate to erythrocytes. In the first place, as RBCs are destroyed by hemolysis, sequestration, senescence, etc., the ^{51}Cr-labeled hemoglobin released does not tag other unlabeled RBCs. Instead, a portion of the ^{51}Cr is excreted in the urine and some of it is taken up in the liver, spleen, and bone marrow as the tagged hemoglobin is catabolized.

Secondly, a portion of the ^{51}Cr is eluted from the cells. Although the elution is somewhat greater on the day of injection (day 0), by day 1 and thereafter the ^{51}Cr is eluted at a fairly constant rate (about 0.6 to 1.2 percent per day). In addition to the relatively large amount of elution on the day of injection, a sizeable number of RBCs are destroyed at this time. Because of these two phenomena, blood samples are not usually obtained on day 0. Therefore if a normal or near normal survival time is anticipated, oxalated blood samples may be collected, starting on day 1, according to the following schedule:

First week: daily
Second and third weeks: three times a week
Fourth and fifth weeks: twice a week

At the end of this time the radioactivity in each sample is determined.

The gradual daily reduction in radioactivity is due to a combination of both RBC destruction and constant elution of the radiochromate. The decline in radioactivity is approximately exponential, so the radioactive measurements are plotted on semilogarithmic graph paper.

The objective is to find that time at which there has been a 50 percent reduction in radioactivity, or $T_{1/2}$. The data are plotted on the graph, with the ordinate representing radioactivity and the abscissa time in days (Fig. 2). Since the reduction in radioactivity is exponential, a straight line is obtained and is extrapolated to day 0 to determine initial blood radioactivity. The time at which there is a 50 percent fall in radioactivity from this extrapolated value is the $T_{1/2}$, which normally occurs at 26 to 35 days. This is the RBC survival or $T_{1/2}$ value. A ^{51}Cr RBC survival time of 26 to 35 days corresponds to an RBC life span of 120 days.

If a much shorter survival time is anticipated, attempts to determine the

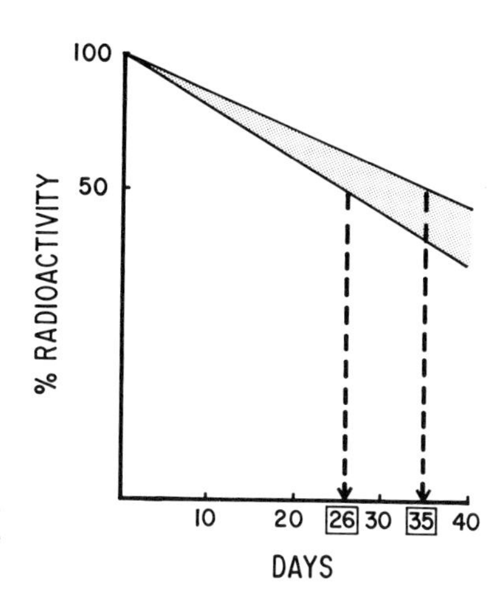

Fig. 2. Erythrocyte survival curve using ^{51}Cr-sodium chromate. The initial decline in radioactivity occurs in exponential fashion, and a 50 percent fall in radioactivity between days 26 and 35 corresponds to a normal erythrocyte life span of approximately 120 days.

$T_{1/2}$ should be made using a smaller number of points. For example, if a patient is suspected of having a rapidly progressing acquired hemolytic anemia, the samples should be counted starting on day 2 in order to obtain a rough estimate of what the survival time might be. In such cases it is usually sufficient to obtain five to seven points for the $T_{1/2}$ determination. Shortened survival times may be seen in such entities as congenital and acquired hemolytic anemias and sickle cell disease.

External Monitoring. Starting on day 0, radioactivity is determined over the heart, liver, and spleen. Counts over the heart are taken as a check on the measurement over the other two organs and for an estimate of blood radioactivity.

The liver/spleen radioactivity ratio is calculated and in the absence of marked splenomegaly is about 1:1 to 1.5:1.0. If the ratio is unchanged for 4 days, further external monitoring probably will not yield more information. On the other hand, if the liver/spleen ratio changes, for example from 1.5:1.0 on day 0 to 1:2 and 1.0:3.8 on the succeeding days there is probably RBC sequestration in the spleen, and attempts may be made to determine the RBC survival time as early as day 3 or 4. If the spleen is unusually large the liver/spleen ratio may be as great as 1:2 on day 0, but in the absence of splenic sequestration of RBCs the ratio will continue unchanged over the following several days of monitoring.

Radioactive DFP Studies. Radioactive DFP is bound to the erythrocyte membrane probably by inhibiting the enzyme cholinesterase. It undergoes elution during the first 7 days after intravenous administration, but thereafter it is firmly bound without further elution. The latter property has made this material attractive to some investigators. However, since the DFP is labeled with ^{3}H or ^{32}P, both beta emitters, radioactive measurements must be made

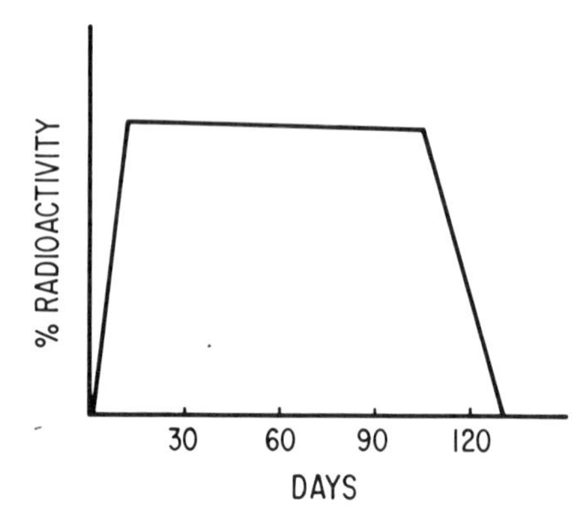

Fig. 3. Erythrocyte life span curve using ^{3}H-glycine. In this study a cohort of erythrocyte precursors are labeled with the tagged glycine.

with a liquid scintillation counter. This makes the procedure more complicated than one involving a gamma emitter (e.g., ^{51}Cr) and in part accounts for the popularity of radiochromate studies.

Cohort Labeling

In the cohort method a radioactive hemoglobin precursor such as glycine labeled with ^{3}H, ^{14}C, or ^{15}N is administered to the patient. The radioactive glycine is incorporated into a cohort of erythrocytes of approximately the same age as they are formed in the marrow. Daily blood samples are obtained and evidence of radioactivity is detected when the labeled cells first appear in the circulation (Fig. 3). The radioactivity rises rapidly, reaches a plateau, and then falls as the labeled cells reach the end of their life span.

Since the tagged RBCs are of approximately the same age, the life span is determined with a fair degree of accuracy as the radioactivity decreases and approaches the base line of the curve. Although this method of measuring the RBC life span is more direct than the random labeling of cells it requires taking daily blood samples for upward of 120 days in the normal individual, thus limiting its clinical usefulness.

FERROKINETICS

The information gained from ferrokinetic studies has greatly amplified our understanding of the physiologic alterations that can occur in abnormalities of erythrocyte production. The data obtained from these studies may be of such a complex nature that the interpretation and analysis of this material should often rest in the hands of a physician especially competent in this area. Much of the work in this field has been summarized in the excellent monograph by Finch et al.[5]

After iron is introduced into the circulation, either through ingestion and absorption or by direct intravenous administration, it is bound to transferrin.[6] Circulating as a transferrin-iron complex, it leaves the plasma rapidly to enter the erythrocyte precursors. In this process the transferrin-iron complex becomes attached to the erythroid cell membrane and the transferrin separates from the complex as the iron enters the cell.

Following the breakdown of both mature and immature erythrocytes, iron is split from hemoglobin within the reticuloendothelial (RE) cells. Attaching to transferrin again to circulate as a transferrin-iron complex, the greater portion of this iron derived from catabolized hemoglobin reenters the erythroid marrow and is the principal source of the iron that is incorporated into the hemoglobin of developing cells of the erythroid series. Thus transferrin takes up iron at the RE cell and parts with it at the erythron's membrane.

Ferrokinetic studies are undertaken after intravenous administration of 5 to 10 μCi ^{59}Fe-ferrous citrate. In most cases there is sufficient circulating transferrin present to which the iron may become attached, but if the unsaturated iron binding capacity of the plasma is very low (e.g., in a patient with hemochromatosis) a tagged transferrin-iron complex is prepared by incubating the radioactive iron with another patient's plasma. In such a situation the prepared transferrin-^{59}Fe complex is injected into the patient rather than using ^{59}Fe-ferrous citrate directly.

The immediate objectives of ferrokinetic procedures are to gain information in regard to the following:

1. The rapidity with which radioactive iron leaves the plasma (through analysis of plasma samples). Normally ^{59}Fe disappears from the plasma as it enters the erythroid marrow.
2. The site or sites of radioiron accumulation (through external monitoring).
3. The time required for incorporation of the radioiron into circulating erythrocytes (through analysis of blood samples).

Plasma Disappearance

To study the plasma disappearance of radioactive iron, the ^{59}Fe-ferrous citrate is injected intravenously and plasma samples are obtained at intervals of 5, 15, 30, 60, 90, and 120 minutes after tracer administration. Plasma radioactivity in each sample is determined, and since the reduction in ^{59}Fe radioactivity proceeds in exponential fashion the data are plotted on semilogarithmic coordinates (Fig. 4). The objective is to determine the $T_{1/2}$, or the time at which there has been a 50 percent fall in radioactivity. The activity at time 0 may be obtained by extrapolation to the ordinate and the $T_{1/2}$ readily determined; normally it is 60 to 100 minutes.

The avidity of the erythroid marrow for iron is increased in such conditions as iron deficiency anemia and polycythemia vera; here the $T_{1/2}$ is shorter than normal. On the other hand, if there has been extensive marrow destruction or replacement, as in aplastic anemia, the $T_{1/2}$ is prolonged.

It should be remembered that in the normal subject approximately 40 percent of the radioiron that entered the marrow returns to the plasma several hours or days after tracer administration. This is known as reflux iron.

A normal plasma disappearance curve in itself does not always signify that radioiron is entering the marrow to become incorporated into the hemoglobin of normally developing erythrocytes. For example, ^{59}Fe may disappear from the plasma rapidly and enter structures other than the marrow, such as the hepatic parenchymal cells. Even if the radioactive iron does enter the erythroid marrow, there is no indication from the normal plasma disappearance curve whether such uptake will result in viable erythrocytes or in those destined to undergo destruction while still in the erythroid marrow (ineffective erythropoiesis). Such questions are clarified in part by further

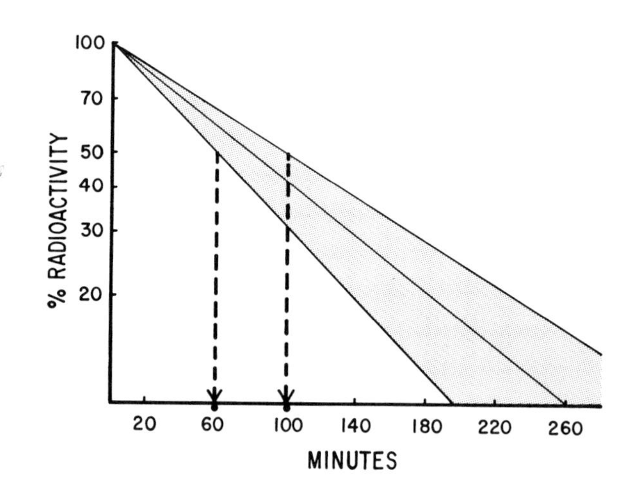

Fig. 4. Plasma disappearance of radioactive iron after ^{59}Fe-ferrous citrate administration. The decline in radioactivity is exponential as the radioiron leaves the plasma to enter the reticuloendothelial cells in the bone marrow. Normally there is a 50 percent fall in radioactivity 60 to 100 minutes after ^{59}Fe injection.

ferrokinetic studies, such as external organ monitoring and analysis of iron incorporation into erythrocytes.

External Monitoring

There are three structures in which relatively large amounts of iron may be stored: erythrons, RE cells, and the parenchymal cells of the liver.[5] Iron enters the erythrocyte precursor as the transferrin-iron complex becomes attached to the cell membrane in the erythroid marrow. RE cells can take up only that iron released from erythroid cells, both mature and immature, that have undergone destruction. Although such uptake normally occurs only in the RE cells of the erythroid marrow, iron from erythrocyte breakdown may also be seen in RE cells elsewhere (e.g., Kupffer cells of the liver and splenic macrophages) in certain pathologic states. On the other hand, the liver's parenchymal cells can take up transferrin-iron as such; separation of the iron from transferrin is not necessary with these cells.

To determine the site or sites of radioiron accumulation, external monitoring is performed over the precordium, liver, spleen, and sacrum. The latter reflects marrow activity, while the precordial counts are indicative of blood radioactivity.

Organ counting may commence 1 hour after tracer injection, and measurements are obtained at hourly intervals up to 6 hours. Thereafter organ counting is performed once daily for the next 10 to 12 days. The day on which the radioiron is administered is referred to as day 0.

Normally the marrow radioactivity shows an initial rise and reaches a peak between days 1 and 2 (Fig. 5). As newly formed erythrocytes containing radioiron are discharged from the marrow into the circulation, the marrow radioactivity declines. Although the blood radioactivity falls initially as the radioactive iron enters the erythroid marrow, its radioactivity gradually increases with the appearance in the bloodstream of erythrocytes into whose hemoglobin ^{59}Fe has become incorporated. Liver and spleen counts in the normal individual reflect somewhat the blood radioactivity.

Abnormalities in Splenic Radioactivity. A gradual rise in splenic radioactivity following the initial decline in blood radioactivity is indicative of splenic hematopoiesis (Fig. 6). On the other hand, a sharp rise in the counts over the spleen after the marrow radioactivity has risen and begun to fall suggests that erythrocytes are undergoing destruction or sequestration in the spleen.

Abnormalities in Hepatic Radioactivity. If hepatic uptake shows an initial and continued rise at the same time the blood radioactivity is decreasing, there is probably no hematopoietic activity in the marrow or elsewhere (Fig. 7), as in aplastic anemia. In such a case, hepatic activity is due to the entrance

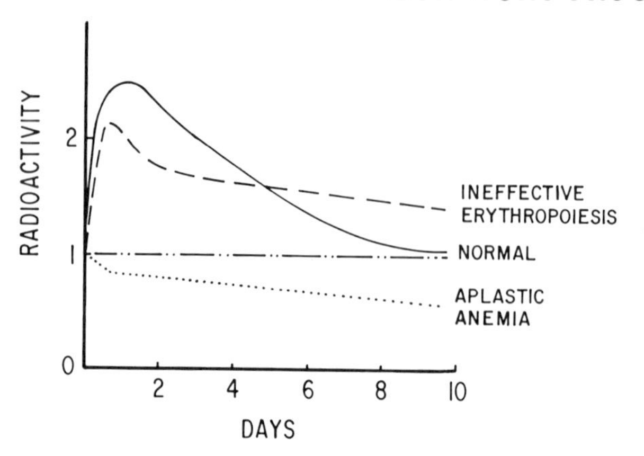

Fig. 5. Bone marrow uptake of radioiron. Radioactivity is presented in terms of relative units, with 1 representing the organ or tissue radioactivity immediately after injection of ^{59}Fe. Normal: As the injected radioactive iron leaves the plasma there is an associated rise in marrow radioactivity. This activity falls as those newly formed erythrocytes, into which radioiron is incorporated, are discharged into the bloodstream. Such a rapid decline in radioactivity is not seen with ineffective erythropoiesis since erythrocytes are destroyed in the marrow before they can be discharged to the circulation. Since no radioiron enters the marrow in aplastic anemia, the activity in such areas declines. (After Finch et al.[5])

of transferrin-iron into the parenchymal cells of the liver. On the other hand, extramedullary hematopoiesis in the liver may be associated with a rise in hepatic activity following the initial decline in blood radioactivity, findings similar to those seen in splenic hematopoiesis. The uptake in this instance is in the RE cells as the radioiron enters the endothelial lining Kupffer cells.

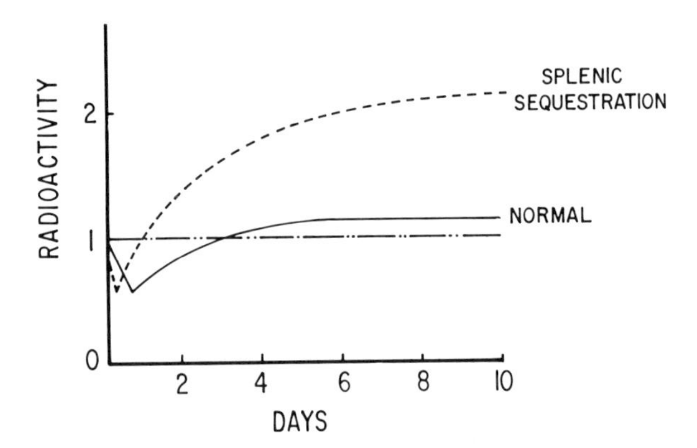

Fig. 6. Splenic uptake of radioiron. Splenic sequestration: As erythrocytes containing the radioactive iron are formed in the marrow and are subsequently discharged to the circulating blood, they become trapped in the spleen and the latter's radioactivity exhibits a rise. (After Finch et al.[5])

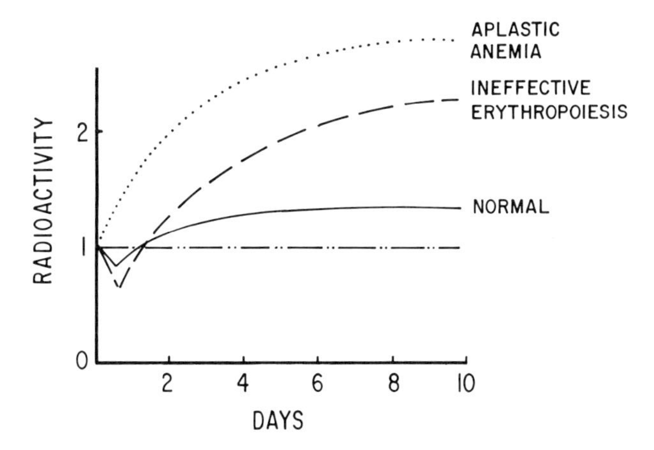

Fig. 7. Hepatic uptake of radioiron. Aplastic anemia: The parenchymal cells of the liver are able to take up the radioiron directly as transferrin-iron. Ineffective erythropoiesis: After the initial decline in radioactivity as the ^{59}Fe enters the marrow, there is a gradual rise in radioactivity as the nonviable erythrocytes are removed from the circulation by the reticuloendothelial cells, such as the Kupffer cells of the liver. (After Finch et al.[5])

Abnormalities in Marrow Radioactivity. Virtually no marrow radioactivity is recorded in aplastic anemia. With ineffective erythropoiesis in which radioiron becomes incorporated into the hemoglobin of erythroid cells that are destroyed while still in the marrow, there is normal initial uptake of radioiron. However, the radioactivity in the bone marrow remains relatively high. This is because much of the iron released as the erythrocyte precursors are destroyed is taken up in other developing erythroid cells and because so few erythrocytes are discharged by the marrow to the bloodstream (Fig. 5).

Incorporation of Iron into Erythrocytes

After the radioactive iron enters the erythroid marrow, it is incorporated into the hemoglobin of developing erythrocytes over the next several hours. A small number of erythrocytes containing ^{59}Fe are discharged into the circulation during days 1 and 2. However, following this there is a continual release of radioiron-containing erythrocytes over a period of about 2 weeks, this discharge occuring in exponential fashion. Normally about 80 percent of the radioactive iron has become incorporated into circulating erythrocytes 12 to 14 days after ^{59}Fe administration.

Samples of whole blood are obtained daily from days 1 through 14. The daily percent incorporation of ^{59}Fe into erythrocytes may be determined from the following equation:

$$\% \text{ RBC utilization} = \frac{\text{cpm/ml whole blood} \times \text{blood volume} \times 100}{\text{cpm administered } ^{59}\text{Fe}}$$

Since the radioactivity in the plasma at zero time can be obtained from the extrapolated value in Figure 4, the plasma volume and whole blood volume may be calculated:

$$\text{Plasma volume} = \frac{\text{cpm administered } ^{59}\text{Fe}}{\text{cpm/ml plasma at time 0}}$$

$$\text{Blood volume} = \frac{\text{Plasma volume}}{\text{corrected hematocrit}}$$

The iron utilization curve in a patient with iron deficiency anemia shows an elevated RBC incorporation of iron, with maximum incorporation occurring earlier than normal (Fig. 8). Maximum incorporation also occurs early in a patient with a rapidly progressing hemolytic anemia, but here the degree of incorporation is usually well below normal, as the erythrocytes undergo destruction. If a large portion of the developing erythrocytes undergo

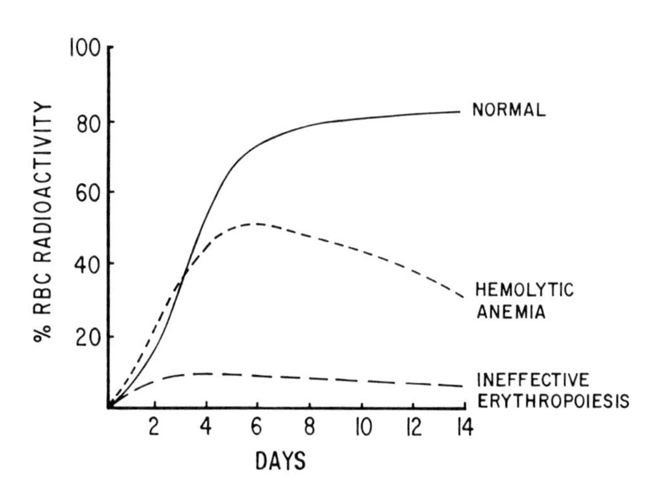

Fig. 8. Erythrocyte utilization of radioiron (incorporation of iron into the cells). Normal: There is maximum incorporation of ^{59}Fe into circulating cells 2 weeks after injection. Hemolytic anemia: A highly active marrow is associated with early appearance of the erythrocytes in the bloodstream; the reduction in radioactivity is dependent on the extent and time of erythrocyte destruction. Ineffective erythropoiesis: Few ^{59}Fe-containing erythrocytes reach the peripheral circulation. (After Finch et al.[5])

destruction in the marrow before they can enter the bloodstream (ineffective erythropoiesis), only scant radioactivity is detected.

SCHILLING TEST

The Schilling test has been employed for the past 20 years to assist in the diagnosis of pernicious anemia.[7] The study involves the degree to which a labeled sample of vitamin B_{12} (cyanocobalamin) is absorbed from the small intestine and subsequently excreted in the urine, where its radioactivity can be easily measured.

Intrinsic factor (IF), which is necessary for vitamin B_{12} absorption, is secreted by the parietal cells of the gastric mucosa. Such IF secretion does not occur in pernicious anemia. Although the mechanism of absorption is not yet fully understood, the IF-B_{12} complex probably becomes attached to a receptor cell in the villous lining of the ileum. As the complex passes through the intestinal wall the vitamin B_{12}, after separating from the IF, enters the portal blood where it becomes attached to vitamin B_{12}-binding proteins (transcobalamins) and is rapidly delivered to the parenchymal cells of the liver.[8] Although vitamin B_{12} enters other tissue cells too, its principal binding sites are the circulating transcobalamins and hepatic cells.

When performing the Schilling test, the patient is instructed to fast after midnight and is given 0.5 μCi ^{57}Co-cyanocobalamin by mouth in the morning. (This contains approximately 0.5 μg vitamin B_{12}.) A 24-hour urine collection is started at the time the radionuclide is given. Two hours later the patient receives an intramuscular injection of 1000 μg nonradioactive vitamin B_{12}, the so-called flushing dose used to "saturate" the hepatic and plasma vitamin B_{12}-binding sites. Absorption of the ^{57}Co-vitamin B_{12} commences about 3 to 4 hours after ingestion of the radioactive material, so by the time it enters the bloodstream no free vitamin B_{12}-binding sites are available. Thus the absorbed ^{57}Co-cyanocobalamin must bypass the overloaded transcobalamins and hepatic cells and is excreted in the urine.

The radioactivity in the urine is determined by obtaining the counts per minute per milliliter of a sample and then multiplying this by the urine volume. By comparing the radioactivity in the 24-hour urine collection with the radioactivity in the administered dose, the percent excretion can be determined:

$$\% \text{ excretion} = \frac{\text{cpm/ml urine} \times \text{24-hour urine volume(ml)}}{\text{cpm administered } ^{57}\text{Co-vitamin B}_{12}} \times 100$$

Normally the excretion is 7 percent or more in 24 hours. If it is less than that, the possibility of pernicious anemia is considered and the test should be repeated no sooner than 7 days later with the addition of 60 mg IF (hog gastric mucosa). If excretion is then 7 percent or greater with the IF the findings are consistent with pernicious anemia. If the values continue to be less than 7 percent, a cause other than the absence of IF should be considered, such as malabsorption.[9]

One of the problems with the Schilling test is the difficulty in obtaining accurate 24-hour urine collections in the hospitalized patient. To help eliminate this problem, studies to determine plasma ^{57}Co-vitamin B_{12} levels have been undertaken. Although such a procedure is technically more difficult, results have been reliable in distinguishing the normal subject from the patient with reduced vitamin B_{12} absorption.

FAT ABSORPTION STUDIES

The evaluation of patients with steatorrhea is now commonly performed in the nuclear medicine laboratory by determining the degree to which a radioiodinated neutral fat or fatty acid is absorbed after ingestion.[10] The popularity of these radionuclide examinations is a testimonial to both the reliability of the studies and the satisfaction among laboratory personnel when called upon to analyze blood radioactivity rather than stool specimens.

Before neutral fats or triglycerides can be absorbed, they must first be hydrolyzed by pancreatic lipase into free fatty acids and glycerol. It is in the form of free fatty acids that intestinal absorption principally occurs, although a minor amount of triglycerides may be absorbed directly (no more than 10 percent).

In studying a patient with steatorrhea for the presence or absence of malabsorption, the patient is given a fat-free meal the evening before the examination and receives nothing by mouth after midnight. The following morning 50 μCi ^{131}I-oleic acid is administered by mouth. To determine the extent to which the radioiodinated fatty acid is absorbed, whole blood samples are withdrawn 2, 4, and 6 hours after tracer ingestion. The radioactivity of each sample (counts per minute per milliliter) is obtained, and the total blood radioactivity is calculated by multiplying this value by the estimated blood volume (obtained through a standard blood volume chart according to the patient's height, weight, and sex). The percentage absorption of radionuclide for each given time is then determined by comparing the total blood radioactivity with the radioactivity present in the administered sample of ^{131}I-oleic acid.

$$\%\ \text{absorption} = \frac{\text{cpm/ml whole blood} \times \text{blood volume in ml}}{\text{cpm } ^{131}\text{I-oleic acid administered}} \times 100$$

The values are then plotted on a graph (Fig. 9). In a typically normal study such as that in Figure 9, the peak radioactivity of approximately 14 percent may occur about 4 hours after tracer administration; thereafter the radioactivity may decline slightly.

There is no general agreement about the times at which blood samples are obtained and normal values of the absorption percentage of radioactive material are established. Some institutions prefer to obtain, for example, hourly samples for 6 hours or specimens at 1, 3, 5 and 6 hours. Peak values as low as 9 percent are considered normal in some laboratories, while in others the study is deemed normal when the average of the samples at 4, 5, and 6 hours is 8 percent or more. Obviously it is necessary to establish normal values and procedures in each institution.

If there is failure to absorb the ^{131}I-oleic acid, as in a case of steatorrhea due to malabsorption, there is generally a flattened curve (Fig. 9). The small amount of radioactivity present may be due in part to absorption of a small portion of ^{131}I that has become dissociated from the tagged oleic acid.

Information regarding both pancreatic insufficiency (insofar as the secretion of pancreatic lipase is concerned) and intestinal absorption may be

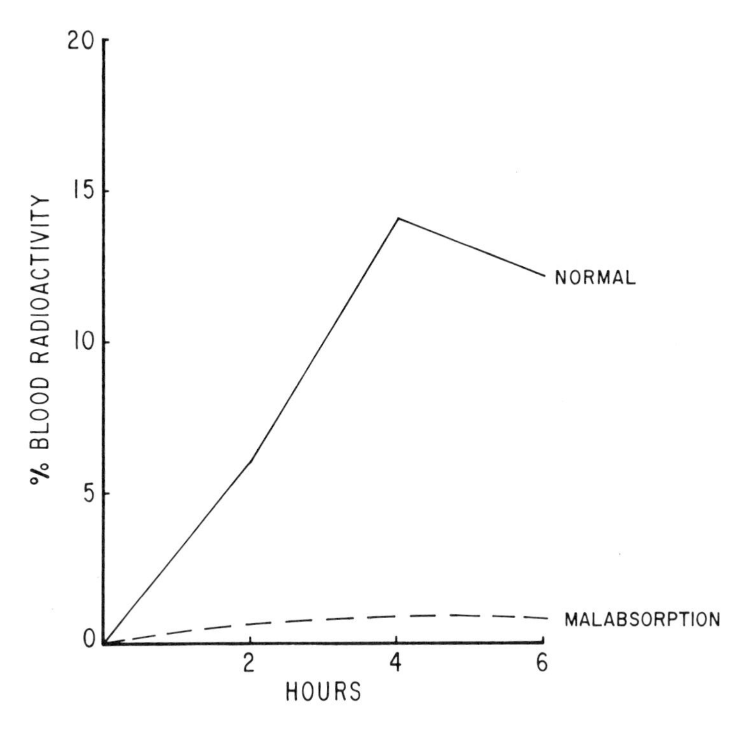

Fig. 9 Blood radioactivity of ^{131}I-oleic acid in a normal subject and in a patient with malabsorption.

obtained with ^{131}I-triolein. For this tagged triglyceride to be absorbed it is necessary that it be hydrolyzed by pancreatic lipase, resulting in free ^{131}I-oleic acid and free glycerol.

The procedure is similar to that for the radioiodinated oleic acid study. Following the fat-free evening meal and nothing by mouth after midnight, the patient receives 50 μCi ^{131}I-triolein the following morning and blood samples are taken 2, 4, and 6 hours later. The radioactivity in each sample is measured, the percent absorption calculated for each of the specimens, and a graph constructed.

Normal values are almost identical to those obtained in the ^{131}I-oleic acid studies, but interpretation is somewhat different. Normal absorption indicates that pancreatic lipase is indeed being secreted to split the ^{131}I-triolein into glycerol and ^{131}I-oleic acid, the radioactivity in the blood being due to the absorbed radioactive oleic acid. If the blood radioactivity is reduced, either pancreatic insufficiency, malabsorption, or both may be present.

Approaches vary as to whether the steatorrheic patient undergoing absorption studies should be tested with radioactive triolein or oleic acid first. If malabsorption is the primary consideration, the patient should be studied with ^{131}I-oleic acid. A normal oleic acid study indicates that malabsorption is not present and may suggest the possibility that the steatorrhea is associated with pancreatic lipase deficiency. This may be confirmed if blood radioactivity is abnormally low aftter the ^{131}I-triolein test. It should be noted that the radioactive absorption studies should be performed no sooner than 1 week apart in order to reduce residual blood radioactivity, and of course the latter must be subtracted from the blood samples in the subsequent study.

Others prefer to start the investigation with radioiodinated triolein. If the triolein test shows reduced absorption, administration of ^{131}I-oleic acid may offer further clarification: A normal oleic acid test points to pancreatic insufficiency, while an abnormal one is indicative of intestinal malabsorption, without being able to shed further light on whether there is a deficiency of pancreatic lipase.

RADIOASSAY

The development of in vitro methods to measure the concentration of body substances with radioactive materials has progressed to an ususual degree during the past decade. These radioassay procedures were introduced and perfected principally by several outstanding investigators working independently in the United States, Britain, and Canada. In the United States progress in this area is most closely linked to the ingenious work of the late Dr. Solomon A. Berson.

Suppose the concentration in the blood of a particular substance, X, is to be determined.[11] It is common to refer to this substance as the *native ligand.* The following are then obtained:

1. A sample of substance X (from a source other than the patient), which is labeled with a radioactive material, such as ^{131}I, and which may then be indicated as $^{131}I\text{-}X$. This is known as the *labeled* or *tracer ligand.*
2. An agent to which the ligands, both native and labeled, will become bound and which may be designated as Y.

If the tracer ligrand is added to the binding agent, the complex $^{131}I\text{-}X \bullet Y$ results.

If the unlabeled or native ligand is introduced into the system, it competes with the tracer ligand for the binding agent. The native ligand displaces the tracer ligand from the $^{131}I\text{-}X \bullet Y$ complex or inhibits the binding of the tracer ligand in proportion to the concentration of native ligand in the system (Fig. 10).

The free or displaced $^{131}I\text{-}X$ may be separated from the bound substance by one of several methods—e.g., by use of a resin or by paper chromatography. The radioactivity for both the bound and free portions may then be determined.

In conjunction with the initial determinations with the unknown concentration of native ligand, the procedure is duplicated using the same quantity of tracer ligand and employing preparations containing varying amounts of ligand of known concentrations. Although the data may be expressed in a number of ways, it is common to determine the ratio of bound/free radioactivity (B/F) and plot this information on a graph along with the concentrations in the known specimens (Fig. 11).

The B/F ratio of the unknown sample is then indicated on the curve, and

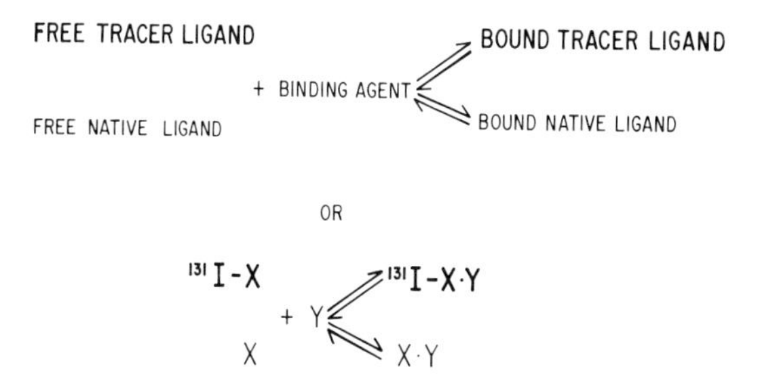

Fig. 10. Competitive binding radioassay. X, native ligand. ^{131}I-X, tracer ligand. Y, binding agent.

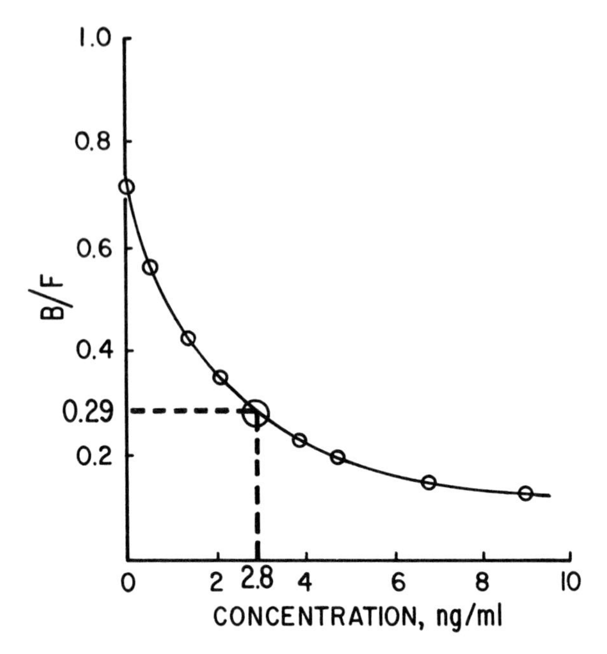

Fig. 11. Competitive binding analysis. The ratio of bound-to-free radioactivity is determined for a number of samples of known concentration, and from this information a curve is constructed. The ratio of bound-to-free radioactivity in the unknown sample is found to be 0.29, and, *from the curve,* the concentration in the unknown sample is found to be 2.8 ng/ml. B/F= ratio of bound-to-free radioactivity; ng = nanograms or millimicrograms.

the unknown concentration of ligand easily determined. As the concentration of native ligand increases, the B/F ratio is diminished (Fig. 11). The concentration of substances measured in such radioassays is minute and is usually expressed in terms of nanograms (millimicrograms or one billionth of a gram) or picograms (micromicrograms or one trillionth of a gram) per milliliter. The data contained in the ordinate may be expressed in terms other than the B/F ratio. Among those commonly used are percent of bound radioactivity and the F/B ratio.

The term competitive binding is probably best applied to radioassays of this type in which the concentration of native ligand is the object of the determination.[12] However, when measuring the plasma concentration of some hormones the binding agent is the antibody to tte hormone. Thus the native and tracer ligands enter into an antigen-antibody complex with the binding agent—hence the term radioimmunoassay.[13]

The brevity with which competitive binding assay and radioimmunoassay are treated in this book is in no way related to the importance of these procedures. These tests are now used extensively, and the availability of commercially prepared kits for competitive binding studies has further simplified them. Although the applications are widespread, the area of greatest attention has been directed to the measurement of hormone concentration by radioimmunoassay. Among the hormones whose concentrations may be determined by these methods are insulin, renin, aldosterone, cortisol, and human growth hormone. Blood levels of such drugs as digoxin may also be determined by radioassay.

These procedures need not be confined solely to measurements of the native ligand. The principle may also be extended to determine the binding capacity of the binding agent for the tracer ligand, this having its greatest clinical application in the determination of the unsaturated binding capacity.

For example, in the determination of unsaturated iron binding capacity in plasma, the iron is firmly bound to the protein transferrin. If any binding sites for iron in the transferrin are not yet occupied, the degree to which these unbound sites are present is an expression of the unsaturated iron binding capacity. If the transferrin iron level is low and the occupied sites are few, the unsaturated iron binding capacity is high. The T3 resin uptake is, in a sense, a measure of the unsaturated binding capacity of plasma proteins for thyroid hormone.

AUTORADIOGRAPHY

Suppose an animal receives an intravenous injection of a radioactive material and a microscopic section is obtained later from an organ or tissue in which the radionuclide has accumulated. A slide containing the histologic section is then dipped into a liquid photographic emulsion. The radioactivity in the tissue section causes reactions to occur in the photographic emulsion so that, after exposure, developing, and fixing, the site of radionuclide localization in the tissue can be determined. To better understand the phenomena involved, it is helpful to consider the action of light on a photographic emulsion. [14]

A photographic emulsion is composed of a latticework of silver bromide crystals suspended in gelatin. The imperfections in the latticework, or irregularities, are known as sensitivity specks. When a light photon strikes the emulsion an electron is liberated from a bromide ion. This electron migrates to the sensitivity speck where it becomes trapped and causes the positively charged silver ion to be converted into an atom of metallic silver.

The deposition of several atoms of metallic silver at the sensitivity speck is responsible for what is known as a latent image.

A photographic developing solution acts as a reducing agent, causing silver ions to be reduced to metallic silver. If the emulsion that was exposed to the light photon is later immersed in developer, the metallic silver already formed at the sensitivity speck acts as a catalyst for the deposition of more metallic silver at this site. To prevent all the silver bromide from becoming converted to silver atoms, developing is stopped after a short time (2 to 3 minutes) by adding distilled water or a weak acetic acid solution. The developed emulsion is then placed in fixer resulting in the silver bromide crystals being dissolved via the formation of silver complexes; the remaining silver bromide is thus brought into solution. The pattern or configuration of the silver grains that remains reflects the pattern of the light photon to which the emulsion had been exposed.

The light photon affects the photographic emulsion in a manner analogous to ionizing radiation. If, for example, a beta emitter comes in contact with the emulsion, this negatively charged particle may collide with an electron in the outer orbit of a bromide ion. The liberated electron can then travel to a sensitivity speck where the silver ion is converted to an atom of metallic silver and a latent image is thus produced at the site at which the radioactivity is present. This, then, is the basic reaction involved in autoradiography in which silver grains are produced.

Of the ionizing radiations that have been employed in autoradiography, beta emitters are by far used most frequently. Few alpha emitters have been involved in medical research.

It is also possible to perform autoradiography utilizing other types of ionizing radiation, such as internal conversion electrons. Some isotopes that are relatively weak gamma emitters, (e.g., ^{125}I and ^{99m}Tc) decay also by internal conversion. These electrons behave similarly to beta particles in the photographic emulsion. Since the likelihood of a gamma ray colliding with an orbital electron in the emulsion is so remote, such non-ionizing radiation is not useful in autoradiography.

There are a number of methods available for producing autoradiographs. The following briefly described procedure is only one of many that can be utilized. After radionuclide injection a portion of the tissue to be studied is placed in a fixative that does not affect the emulsion (e.g., Carnoy's solution). The tissue then undergoes dehydration and paraffination, and sections approximately 5 μ in thickness are placed on a gelatin-coated slide. The latter acts as an adhesive suface for both the tissue section and emulsion. The tissue section then undergoes deparaffination.

The slides are dipped in a liquid emulsion for about 5 seconds in either total darkness or using a 15 watt bulb enclosed in a Wratten No. 3 filter. The slides are allowed to dry, placed in light-tight boxes containing a desiccant, and transferred to a refrigerator so that the tissue section containing the radioactive material may be exposed to the emulsion for a number of days, weeks, or even months. (The specimens are kept at about 4 C because heat may cause reactions in the emulsion.) When exposure has been sufficient the slides are ready to be developed, again in total darkness or using a No. 3 Wratten filter. The exposed slides are taken through solutions of developer, fixer, and water, and then undergo histologic staining with hematoxylin and eosin.

An autoradiograph is shown in Figure 12. The radioactivity in the section is indicated by the dark, silver granules overlying the tumor cells.

Numerous other methods for performing autoradiography are widely used. For gross information, a large section of an organ may be placed on a photographic plate. Some investigators may prefer to use stripping film rather than liquid emulsions for examination under the light microscope. Recent innovations include autoradiography with the electron microscope and preparation of sspecimens with a freeze-dry technique in those instances in which the radionuclide may be soluble or diffusible in liquid fixatives.

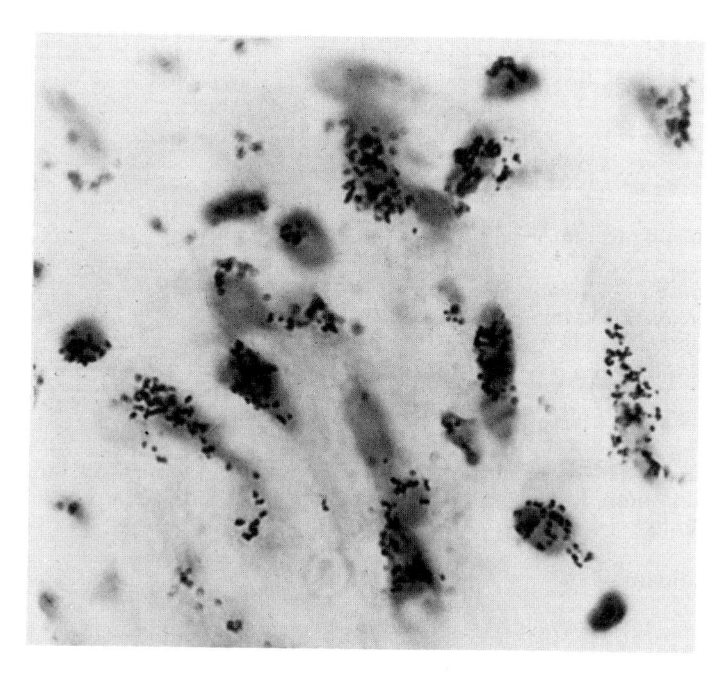

Fig. 12. Autoradiograph showing the localization of ^{99m}Tc-sodium pertechnetate (dark, silver granules) in the tumor cells of a human acoustic neuroma.

REFERENCES

Blood Volume

1. Powsner ER, Raeside DE: Diagnostic Nuclear Medicine. New York, Grune & Stratton, 1971, pp 293-339
2. Silver S: Radioactive Nuclides in Medicine and Biology, 3rd ed. Philadelphia, Lea & Febiger, 1968, pp 271-282
3. Paulsen FR: The determination of blood volume. In Szirmai E (ed.): Nuclear Hematology. New York, Academic Press, 1965, pp 133-170

Erythrocyte (Red Blood Cell) Survival

4. Berlin NI: Life span of the red cell. In Bishop CW, Surgenor DM (eds.): The Red Blood Cell. New York, Academic Press, 1964, pp 423-450

Ferrokinetics

5. Finch CA, Deubelbleiss K, Cook JD, et al: Ferrokinetics in man. Medicine (Baltimore) 49:17, 1970
6. Fairbanks VF, Fahey JL, Beutler E: Clinical Disorders of Iron Metabolism, 2nd ed. New York, Grune & Stratton, 1971, pp 42-127

Schilling Test

7. Schilling RF: Intrinsic factor studies. II. The effect of gastic juice on the urinary excretion of radioactivity after the oral administration of radioactive vitamin B_{12}. J Lab Clin Med 42:860, 1953
8. Beck WS: Vitamin B_{12} deficiency. In Williams W, Beutler E, Erslev AJ, et al (eds): Hematology. New York, McGraw-Hill, 1972, pp 256-278
9. Silberstein E: The Schilling test. JAMA 208:2325, 1969

Fat Absorption Studies

10. Kalser MH: Tests of malabsorption syndrome. In Bockus HL (ed.): Gastroenterology, 2nd ed. Philadelphia, Saunders, 1964, Vol 3, pp 492-497

Radioassay

11. Yalow RS, Berson SA: General principles of radioimmunoassay. In Hayes RL, Goswitz FA, Murphy BAP (eds.): Radioisotopes in Medicine: In Vitro Studies. Oak Ridge,Tenn., US Atomic Energy Commission, 1967, pp 7-41
12. Murphy BAP: Protein binding and its application in the assay of nonantigenic hormones: thyroxine and steroids. In Hayes RL, Goswitz FA, Murphy BAP (eds.): Radioisotopes in Medicine: In Vitro Studies. Oak Ridge, Tenn., US Atomic Energy Commission, 1967, pp 43-57

13. Berson SA, Yalow RS: Principles of immunoassay of peptide hormones in plasma. In Astwood EB, Cassidy CE (eds.): Clinical Endocrinology, 2nd ed. New York, Grune & Stratton, 1968, pp 699-720

Autoradiography

14. Rogers AW: Techniques of Autoradiography. Amsterdam, Elsevier, 1967, pp 10-48

PART THREE

Physical Aspects of Nuclear Medicine

chapter 8

RADIONUCLIDES: CHARACTERISTICS AND PRODUCTION

In order to use radionuclides effectively as a diagnostic aid their physical nature must be understood. A detailed knowledge of nuclear physics is not necessary, although it is essential to understand the basic concepts of nuclear structure and nuclear reactions.

In a physical sense the material world around us is made up of matter that occupies space and has weight. Any material object may be broken into smaller and smaller portions. If this process is continued, a point is reached when the smallest quantity of matter is present that can exist by itself while retaining all the properties of the original substance. This minute quantity of matter is designated a molecule. Molecules may be broken down by chemical action into smaller and smaller elemental units called atoms, which can not be broken down by ordinary chemical changes or be made by chemical union. Thus the smallest part of an element that can participate in a chemical change is the atom.

ATOMIC STRUCTURE

Under certain circumstances changes may be brought about within the atoms of almost any element. Such subatomic changes involve the particles that make up the nucleus or central core of the atom. The energies of these subatomic (nuclear) reactions—far beyond those associated with ordinary chemical reactions—are what make the techniques of nuclear medicine possible. To understand these high energy nuclear reactions the structure of the atom must be explored.

Ninety-two elemental types of atoms exist in nature. The atoms of each elemental species have a characteristic structure composed of various combinations of particles smaller than the atom. The presence and characteristics of these particles has largely been determined by disrupting various types of atoms with the aid of high energy particle accelerators such as the cyclotron. The atoms of individual elements consist of various combinations of three subatomic particles: protons, neutrons, and electrons.[1] In general atoms may be likened to submicroscopic solar systems. There is a central core or nucleus composed of a closely packed group of protons and neutrons. Around this nucleus electrons revolve in rather indistinct orbits at a relatively great distance. These orbits are somewhat analogous to the orbits of the planets about the sun (Fig. 1).

The chemical behavior of an element is largely determined by the number of electrons in the outermost orbit. The significant properties of radioisotopes, or radionuclides, are determined by the structure of the nucleus or core of the atom. The atom may be better understood by considering the properties of the particles of which it is composed.

The *electron* is an elementary particle consisting of a charge of negative electricity. It is the easiest subatomic particle to isolate, being released from some of the metals by so simple a process as heating. The mass or weight of an electron is 9.1×10^{-28}g or about 1/1840 of the mass of one hydrogen atom. The electron has one unit of electrical charge of negative sign. This negative electrical charge makes the electron easily detectable. In relation to

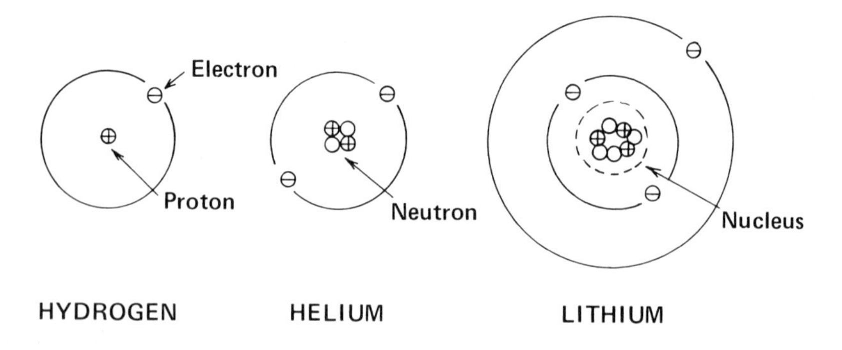

Fig. 1. Structure of the first three atoms of the periodic table of elements.

its mass the electron has the highest electrical charge of any known particle. This enormous charge/mass ratio makes the electron extremely responsive to electrical fields. Comparatively weak electrical fields accelerate an electron to enormous velocities. Because of their electrical charge, electrons in motion interact strongly with atoms nearby. This interaction allows moving electrons to be easily halted by material barriers placed in their path. The high mobility of the electron under the influence of modest electrical fields makes it useful in radio, television, and the many electronic devices common in our society.

The *proton* is an elementary particle identical to the nucleus of a hydrogen atom. Protons are present in the nuclei of all atoms and may be isolated through the ionization of hydrogen atoms of mass number one. The proton has a mass 1840 times that of the electron. The electrical charge on a proton is equal in magnitude to that of the electron, but is of positive sign. Although of considerable mass, the moving proton is easily halted by material barriers in its path due to its relatively large size and its electrical charge. The positive electrical charge makes the proton easily detectable. Protons are sometimes employed as the projectile in cyclotrons because of their significant mass.

The *neutron* is an elementary particle very much like the proton. The most significant difference between these two particles is that the neutron lacks an electrical charge; also its mass is slightly greater than that of the proton. Not easily isolated, neutrons may be liberated from suitable target materials by bombardment with high speed protons or other heavy particles in a cyclotron. Large quantities of neutrons are obtained from the fission of uranium in a nuclear reactor. The absence of electrical charge makes the neutron difficult to detect and when in motion hard to stop. Since neutrons are uncharged particles they can enter the inner nuclear structure of atoms much more readily than charged particles (protons or electrons). This ability to penetrate atomic nuclei is of great significance in the production of radioactive isotopes. Unlike electrons and protons, neutrons have a limited life outside the atomic nucleus. Within a few minutes after ejection from an atom a neutron undergoes spontaneous disintegration giving rise to a proton and an electron. Thus the neutron is composed of a combination of an electron and a proton, and is a stable entity only in the atomic nucleus. Large quantities of neutrons are obtained through controlled splitting of uranium atoms in nuclear reactors; this process yields enormous amounts of energy that can be used for generation of electrical power. The neutrons released in the reactor may also be used to bombard a wide variety of elements for the production of radioisotopes.

Atoms of the various elements are composed of known arrangements of protons, neutrons, and electrons. The simplest element is hydrogen, whose nucleus consists of one proton. The number of neutrons in the nucleus of a hydrogen atom may vary from zero to two; common hydrogen has no neutrons. In fact hydrogen is unique in being the only atom able to exist without neutrons; all atoms other than hydrogen have neutrons in all configurations of the nucleus. All hydrogen atoms have one—and only one—proton in the nucleus. This proton is balanced by an outer orbital electron. The number of electrons in the outer orbits of a given atom are always equal to the number of protons in the nucleus. The nucleus has a net positive charge due to the protons. Since the number of negative electrons in the outer orbits equals the number of protons in the nucleus of a given atom, the atom as a whole is electrically neutral. Starting with hydrogen, which has one nuclear proton, each element of the atomic table has a different number of protons. The most complex element in nature is uranium, which has 92 protons in the nucleus and 92 electrons in outer orbits. In the final analysis, it is the number of protons in the nucleus which gives an atom its identity.

An important characteristic of an atom is the *atomic number,* which is equal to the number of protons in the nucleus. A second important characteristic is the *mass number,* or *atomic weight.* The *mass number* is the sum of the weights of the protons plus the weights of the neutrons in the nucleus of a given atom. It is expressed in terms of multiples of the proton. Each proton and each neutron adds one unit of mass. Thus a hydrogen atom with one proton only has atomic number one and atomic weight one. A hydrogen atom with one proton and one neutron in the nucleus has atomic number one (the number of protons in the nucleus) and atomic mass number two (the number of protons plus the number of neutrons in the nucleus). The atomic number (i.e., the number of protons in a given atom) is often designated by the letter Z. The mass number—the number of protons plus neutrons in a given atom—is represented by the letter A. Hydrogen always has a Z of one, while A may be either one, two, or three depending upon the number of neutrons present. Helium and lithium are located next to hydrogen in the atomic table. Helium has an atomic number of two since the nucleus has two protons. Lithium has an atomic number of three. Helium has a mass number of three or four depending upon the presence of one or two neutrons in the nucleus. Lithium has a mass number of six or seven depending on the presence of three or four neutrons in the nucleus. Nuclear properties of all the other members of the atomic table may be stated in a similar manner. In the shorthand of the chemist, atomic numbers and mass numbers may be included along with the chemical symbol (Fig. 2). The mass number (A) is placed as a superscript to the left of the chemical symbol. The atomic number (Z) is placed as a subscript to the left. In present practice the atomic number is usually omitted. The element in question is designated by its chemical symbol with the mass number in superscript to the left. Thus hydrogen of atomic number one and mass number one is written as ^{1}H. Hydrogen of mass number two, which is also atomic number one, appears as ^{2}H. Helium of

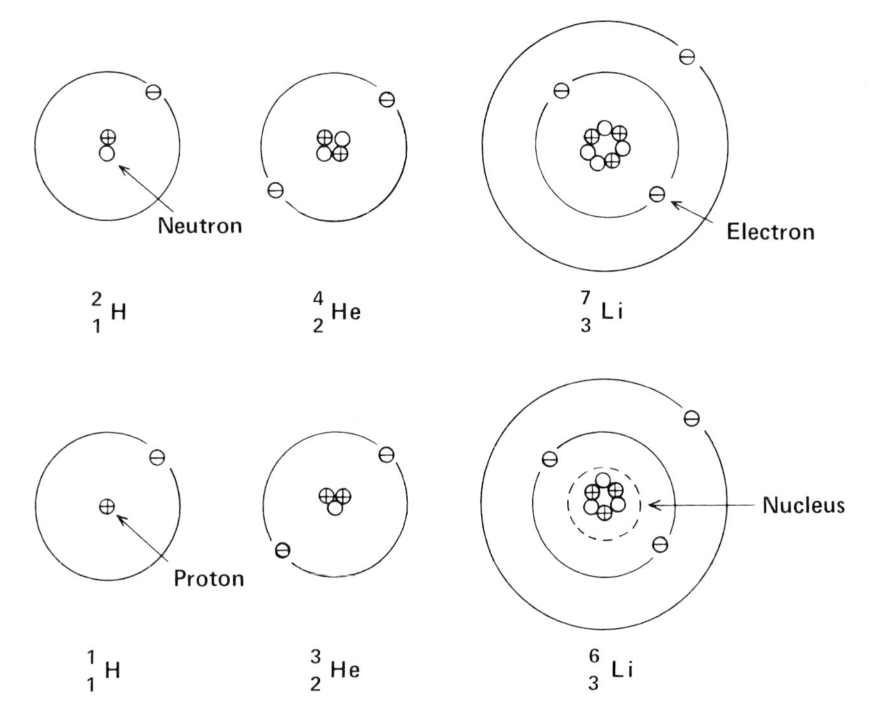

Fig. 2. Chemical symbols and schematic diagrams for two configurations (isotopes) of each of the first three elements of the periodic table.

atomic number two and mass number three is shown as ^{3}He. Lithium with three protons and three neutrons is written as ^{6}Li, and so on through the atomic table.

Isotopes are variations of the same chemical element varying from one another only in the number of neutrons in the nucleus.[2] Thus the nucleus of hydrogen one (^{1}H) has one proton and no other particle. Hydrogen two (^{2}H) has one proton and one neutron in the nucleus (Fig. 2). Hydrogen three (^{3}H) possesses one proton and two neutrons. These three forms of hydrogen are said to be isotopes. The word isotope indicates that they occupy the same place in the atomic table. By this reasoning one isotope of uranium (^{235}U) has 92 protons and 143 neutrons, a total of 235 particles, in its nucleus. A second isotope of uranium (^{238}U) posseses 92 protons and 146 neutrons. The word isotope is often replaced by the more general term nuclide. *Nuclide* is used to designate any species of atom that is characterized by the composition and energy content of the nucleus. Chemical elements isolated from natural sources are often mixtures of isotopes. Tin, for example, has the largest number of naturally occurring isotopes; it has an atomic number of 50 and is found in nature with mass numbers of 112, 114, 115, 117, 118, 119, 120, 122, and 124. The number of protons in a tin atom is fixed at 50, while the number of neutrons may vary from 62 to 74. Each of these various isotopes of tin behaves chemically in an identical way.

Enormous and opposing forces are present within the atomic nucleus. There is a strong binding force tending to hold the nucleus together, while other powerful forces tend to rip it apart. In most cases the forces holding the atom together are continually successful and the atoms are stable. Elements composed of atoms in this group are referred to as stable isotopes. The material world with which we are familiar is made up of stable isotopes of the various elements. There is an interplay between the nuclear forces that depends upon a number of factors, one of which is the ratio of neutrons to protons (Fig. 3). Another is the relative position of the protons and neutrons within the nucleus at any given moment.

If the neutron/proton ratio of a stable atom is changed, the nucleus becomes increasingly unstable. With certain combinations of neutrons and protons the nucleus becomes so unstable that it cannot maintain its structure indefinitely. These atoms are radioactive and comprise the radioisotopes. *Radioactivity* is that property of certain isotopes whereby radiations are emitted as a result of spontaneous changes in the nuclei of atoms of the element.[3] At first glance radioactive atoms do not appear different from their stable counterparts. In time, however, the relative positions of the particles within a radioactive nucleus come into a position of increased disruptive energy that exceeds the stability forces within the atom. When this happens a nuclear reorganization or disintegration takes place. Such disintegrations are accompanied by the emission of radiation. *Radiation* is a process in which energy is emitted from an object, in this case the atomic nucleus, as particles or waves. For a given atom the disintegration point is reached by chance and is unpredictable. For large numbers of atoms the rate of disintegration may be predicted on a statistical basis.

Stable isotopes may be converted to radioactive isotopes by changing the neutron/proton ratio. Additional protons may be added to existing atoms by proton bombardment in a cyclotron. Neutrons may be added by neutron bombardment in a nuclear reactor. Radioisotopes are also created in a nuclear reactor as a result of the fission of uranium atoms, which act as reactor fuel. *Fission* is an unusual type of nuclear transformation that takes place when certain very large atoms are struck by neutrons.[4] In the fission reaction a neutron, upon striking an atom, causes the nucleus to split into two smaller fragments with the release of large amounts of energy. A *nuclear reactor* is a device in which a self-sustaining fission reaction may be initiated and controlled. Uranium 235 is the only naturally available fissionable material. Ar-

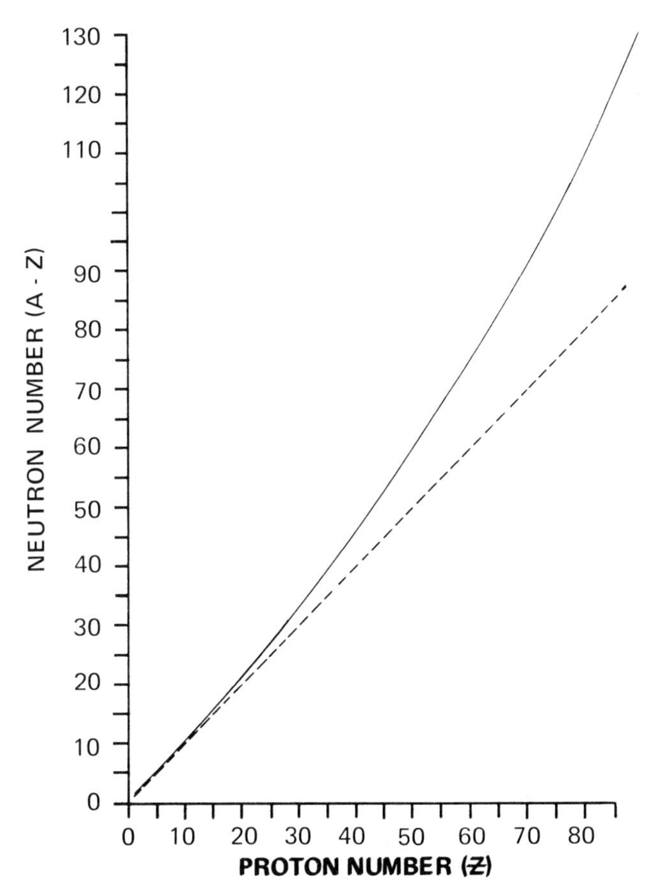

Fig. 3. Neutron/proton ratio for stable isotopes of the various elements. The curved solid line indicates stability. The dotted line represents a 1:1 neutron/proton ratio. Note the greater number of neutrons required for a stable state in the heavier elements.

tificial fissionable materials are made through neutron bombardment of uranium 238 and thorium. The proportion of material in the fission fragments varies from one individual fission to another. Fission fragments for uranium 235 range from element number 30 (zinc) to element 64 (gadolinium). In addition to fission fragments, free neutrons are also obtained in the fission reaction. The free neutrons keep the fission reaction going within the reactor and also provide neutrons for bombardment of materials for radioisotope production. Most of the radioactive isotopes used in medicine are produced in reactors by neutron bombardment. Radioactive cobalt 60 and iodine 131 are examples. Cobalt 60 is a radioactive isotope produced by bombarding stable cobalt 59 atoms with neutrons in a nuclear reactor. One extra neutron is added to each cobalt 59 atom. The single extra neutron is sufficient to make a highly radioactive isotope. Iodine 131 is a radioactive isotope that has four more neutrons than stable iodine 127. Iodine 131 may be obtained from the fission fragments reclaimed from spent reactor fuel.

Once an element is made radioactive its atoms individually undergo a spontaneous internal reorganization *(disintegration)* to reach a stable state. An atom undergoing disintegration is said to *decay*. There are specific modes of decay for given radioisotopes of various elements. Most of the radioisotopes used in medical practice undergo disintegration by *beta emission*. In this process the nuclear neutron/proton ratio is changed by conversion of a neutron into a proton. A neutron within the nucleus changes to a proton by ejection of an electron. The electron is ejected with high velocity, and much energy is carried away from the nucleus by the escaping electron. The electron ejected from the radioactive atomic nucleus is designated a *beta particle*. This ejection is often followed by the emission of one or more bursts of electromagnetic radiation called *photons*. Large amounts of energy may be carried away from the nucleus by electromagnetic radiation. The photons emanating from the nucleus of an atom are called *gamma rays*. Electromagnetic photons of radiation originating outside the nucleus of the atom are designated x-rays or light, depending upon the circumstances of emission and the energy of the photon. All electromagnetic photons coming from the nucleus are designated gamma radiation, regardless of their energy.[5]

Examples of radioactive beta decay are phosphorus 32 and cobalt 60. Phosphorus 32 emits a beta particle only and no gamma radiation. In this spontaneous reaction the atomic mass of the phosphorus remains unchanged at 32 since the number of particles in the nucleus remains the same. As one of the neutrons changes to a proton the atomic number is increased from 15 to 16. This means that the phosphorus of atomic number 15 has changed to a new element of atomic number 16. The new atom is a sulfur atom of atomic number 16 and atomic weight 32. Cobalt 60 (atomic number 27) converts to nickel 60 (atomic number 28) by beta emission. The new nickel nucleus is in an excited state and immediately dissipates its excess energy in the form of two gamma ray photons. Most isotopes decaying by beta emission also emit gamma rays.

A few radioisotopes in medical use decay by electron capture or positron emission. In *electron capture* a proton turns into a neutron by combining with an electron. The electron involved is drawn into the nucleus from the outer orbits (usually the innermost or k orbit) of the atom undergoing the change. In most cases electron capture is accompanied by gamma radiation. In *positron emission* a proton is converted to a neutron by emission of a particle called a positron. A *positron* is identical to an electron except that it carries a positive electrical charge. A positron can exist outside the atomic nucleus for only a short period of time. It soon meets an electron and is annihilated. In *annihilation* a positron combines with an electron, both particles disappear, and two gamma ray photons of specific energy appear at the site of combination.

Chromium 51 and cobalt 58 are examples of isotopes that decay by electron capture. Chromium 51 (atomic number 24) decays to vanadium 51 (atomic number 23) by electron capture; 8 percent of chromium decays are accompanied by gamma ray photons. Cobalt 58 (atomic number 27) decays to iron 58 (atomic number 26); 15 percent of cobalt 58 atoms decay by positron emission and 85 percent by electron capture. All cobalt 58 atoms emit gamma radiation upon decay.

A few heavy radioactive elements, such as radium and thorium, decay by emitting alpha particles. *Alpha particles* are composed of a combination of two protons and two neutrons. Such a combination is immediately recognized as the nucleus of a helium atom. Alpha emitters are not used in medical diagnostic studies.

An important consideration is the energy associated with particles or gamma radiation leaving the nucleus. This energy is usually expressed in terms of electron volts. The *electron volt* is a unit of energy equivalent to the amount of kinetic energy gained by an electron in passing through a potential difference of 1 volt. Care should be taken to differentiate between the electron volt, an energy term, and the volt used in electrical terminology, a potential difference term. The radiations from most radioisotopes have energies amounting to major fractions of a million electron volts. Tables giving the characteristics of various radioisotopes usually express the energy of the emitted radiations in terms of millions of electron volts. A million electron volts is often abbreviated as MeV. For example cobalt 60 yields beta radiation in a continuous range of energies up to 0.306 MeV and specific gamma energies at 1.17 and 1.33 MeV. A knowledge of the energy and type of emitted radiations has several useful functions. An unknown radioisotope may often be identified by measuring the energy of its radiations. Radiation protection considerations are based to a large extent upon the energy of the radiation involved, and the energy and type of radiation must be considered when selecting radiation-measuring instruments. Radiation dosage calculations are also largely based on a knowledge of the energy involved.

DECAY SCHEMES

The disintegration and associated radiations from radionuclides is often represented by a type of schematic drawing which outlines the mode or modes of decay. Energy levels are represented by horizontal lines with the associated energy listed in millions of electron volts. The vertical distance between lines represents the energy difference between nuclear energy levels. Transition between energy levels is represented by arrows according to the following scheme:

1. A vertical arrow pointing downward represents a gamma ray. The energy of the photon is represented by the length of the arrow. There is no change in atomic number.
2. An arrow slanting downward to the right indicates a beta decay. There is an increase of one in atomic number.
3. An arrow slanting downward to the left indicates an electron capture. There is a decrease of one in atomic number.
4. An arrow dropping vertically and then to the left signifies decay by positron emission. The vertical portion of the line is 1.02 MeV long and represents photon energy due to the annihilation of the positron as it combines with an electron. There is a decrease of one in atomic number.

The bottom line is the ground state and the lowest energy level for the nuclide. The ground state is labeled with the elemental symbol, atomic number, atomic mass, and indication of stability (or half-life if not stable). Where various modes of decay exist they may be indicated on the diagram with the percentage of atoms following each path.

Iodine 131, for example, has several modes of decay. The major mode accounts for the decay of 87 percent of all iodine 131 atoms. Examples of *decay scheme diagrams,* constructed as outlined above, are shown in Figure 4. The information contained in the decay scheme may also be presented in tabulated form (Table 1).

Table 1. Decay Scheme: Physical Date for Radioiodine 131

Atomic number (Z)	53
Mass number (A)	131
Half-life (days)	8.05
$\bar{E}_b$(MeV/disintegration)	0.188
Γ(R/mCi/hour at 1 cm)	2.2
β^1(MeV)	0.606
β^2 (MeV)	0.33
Γ^1 (MeV)	0.364
Γ^2 (MeV)	0.638

Note: Ninety percent of the beta and gamma emissions from ^{131}I are included in the above table. The remaining 10 percent includes three additional beta modes and seven additional gamma modes.

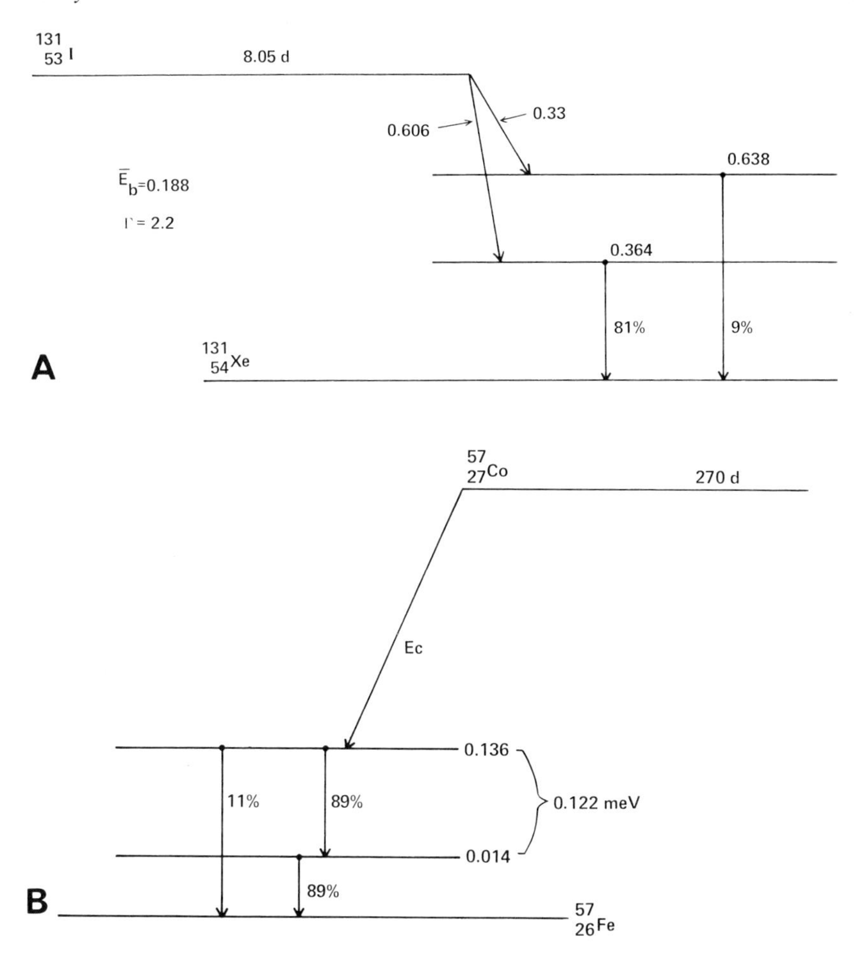

Fig. 4. Examples of decay schemes. A. Two major routes for the decay of iodine 131. B. Complete scheme for the single mode of decay of cobalt 57 (see also notes on p. 247).

The quantity of a radionuclide is measured by the decay rate or the number of atoms disintegrating per second. The basic unit is the curie. One *curie* is equal to 3.7×10^{10} disintegrations per second. The curie is based upon the number of disintegrations per second in 1g radium in equilibrium with its decay products. For most purposes the curie represents an inconveniently large unit. More practical quantities are the *millicurie* (mCi), which is one-thousandth of a curie, and the *microcurie* (μCi), which is one-millionth of a curie. For clinical use radioactive materials are usually purchased in amounts of a few millicuries. Amounts of a few microcuries may be given for some diagnostic studies. Millicurie amounts may be administered for therapeutic purposes.

The quantity of a radioisotope (in curies) may be related to the weight of the material by a factor known as specific activity. *Specific activity* is the total activity per gram or milligram of an isotope. Specific activity may also be expressed as weight per curie or per millicurie. The factor relating weight and activity varies for each isotope. For most radioisotopes the specific activity is such that 1 mCi amounts to a few millionths of a milligram. Gold 198 has a specific activity such that 1 mCi weighs 4.1×10^{-6}mg. Iodine 131 has a weight of 8.1×10^{-6} mg/mCi. Iodine 132 weighs 0.98×10^{-6} mg/mCi. Cobalt 60 weighs 871×10^{-6} mg/mCi. The specific activity of most radioisotopes in clinical use lies within the range associated with these examples.

DECAY CALCULATIONS

The amount of material used in clinical studies is immeasurably small on a weight basis. For instance, the amount of radioiodine necessary for a thyroid study is so insignificant it does not discolor a glass of water or add to its taste. The only practical method for quantitatively measuring a radioisotope is based upon its disintegration rate, which may be determined by detecting the bursts of radiation from disintegrations with a suitable radiation detector.

Unfortunately the use of radioactive materials is complicated by the loss of material by radioactive decay. A correction must always be made for this loss when measuring specified quantities of radiopharmaceuticals. Assume that an iodine antiseptic made with ordinary stable iodine (iodine 127) is purchased. Further assume that it is left sealed and unused upon a shelf. At the end of 1 week the amount of iodine in the container is the same as when first purchased, and 6 weeks later it still remains unchanged. In fact the sample remains unchanged indefinitely. Now assume that a sealed sample of 100 μCi radioactive iodine 131 is purchased and placed upon a shelf, where it is left sealed and undisturbed. At the end of 1 week nearly half of the radioiodine has disappeared, and by the end of 6 weeks only a little over 1 percent remains. As the radioiodine disappears, stable xenon appears in its place. To withdraw a specified amount of iodine 131 from the sample a correction must be made for this loss. The rate of loss by radiodecay varies from isotope to isotope. The loss is usually expressed in terms of the time it takes for half of the quantity of any radioisotope to disappear. This factor, which remains constant for any given isotope, is known as the *half-life*. For example the half-life of iodine 131 is 8.1 days—i.e., if 100 μCi iodine 131 is present at this moment 8.1 days from now one-half or 50 μCi remains, 8.1 days later (16.2 days from now) half of the 50 μCi or 25 μCi remains, 8.1 days after that half of the 25 μCi disappears (so 24.3 days from the starting point only 12.5 μCi remains).This radiodecay continues so that in each succeeding 8.1 days half of whatever is remaining disappears. In theory, the point of complete disappearance of a radioisotope is never reached; in practice, however, the point is reached where the amount remaining is undetectable. Half-lives of various radioisotopes range from a few millionths of a second to billions of years. Most of the radioisotopes used in clinical practice have half-lives of a few hours to a few months. The activity at any time *(t)* may be found for any radioisotope by the equation[6]:

$$At = Ao\ e^{-kt}$$

Ao = activity at the time of initial assay, starting time or time 0
At = activity at the time in question
e = the base of the natural logarithms (2.718)
k = decay constant for the isotope in question
t = time in hours, days, etc., depending upon the half-life of the isotope in question

Since the time may be expressed in hours, days, months, or years, attention must be given to see that the decay constant corresponds to the unit selected for time. The decay constant gives the loss per hour, per day, per month, or per year. The constant must be appropriate to the time interval used. It is interesting to note that the decay constant is related to the half-life ($T_{1/2}$) by the expression:

$$T_{1/2} = \frac{0.693}{k}$$

In verbal form the equation for activity (above) states that the rate of decay of a radioisotope is proportional to the amount present at any given time. Suppliers of radioactive materials usually provide decay tables to aid in calculation. These tables indicate the percentage of material remaining after different time periods following the assay date. This information may also be supplied in graphic form. Semilogarithmic graph paper is used for decay charts. Time is plotted on the linear scale. The fraction of activity remaining is plotted on the logarithmic scale. A decay curve appears as a straight line on semilogarithmic paper. This makes it very easy to construct such a graph if the half-life is known. On ordinary linear graph paper the decay curve is a line of varying curvature that is difficult to plot. Every user of radioactive materials should become proficient in the use of one- and two-cycle semilogarthmic graphs.

The manner in which data on activity and half-life are used is best demonstrated by an example. Consider radioiodinated serum albumin as it is received from the supplier. The container bears the following statements:

^{131}I radioiodinated human serum albumin
Total activity = 1.0 mCi
Quantity = 5.0 ml
Assay = 0.2 mCi/ml
Date of assay = noon, August 19
Expiration date = September 9
Control = 74AS25

The expiration date applies to the biologic properties of the product and has no relation to the decay of the radioiodine. Suppose now that 10 days following assay, on August 29, it is necessary to withdraw 10 μCi from this vial. The volume to be removed may be determined in the following manner: On the assay date the strength was 0.2 mCi or 200 μCi/ml. From decay tables for iodine 131 it is seen that 0.42 or 42 percent of the original iodine remains after 10 days.

$$200\ \mu\text{Ci/ml (initial strength)} \times 0.42 = 84\ \mu\text{Ci/ml (present strength)}$$

$$\text{Volume to be withdrawn} = \frac{\mu\text{Ci to be withdrawn}}{\mu\text{Ci per ml stock solution}} = \frac{10\ \mu\text{Ci}}{84\ \mu\text{Ci/ml}} = 0.12\ \text{ml}$$

REFERENCES

1. Swartz CE: The Fundamental Particles. Reading, Mass, Addison-Wesley, 1965, pp 1-25
2. Friedlander G, Kennedy J: Nuclear and Radiochemistry. New York, Wiley, 1955, pp 29-54
3. Overman RT: Basic Concepts of Nuclear Chemistry. New York, Reinhold, 1963, pp 19-37
4. Stanton L: Basic Medical Radiation Physics. New York, Appleton-Century-Crofts, 1969, pp. 308-328.

5. Hendee WR: Medical Radiation Physics. Chicago, Year Book, 1970, pp 36-39
6. Chase GD, Rabinowitz JL: Principles of Radioisotope Methodology, 3rd ed. Minneapolis, Burgess, 1967, pp 162-169

chapter 9

RADIATION DETECTION

Radioactive materials are dispensed in terms of curies or number of disintegrations per second. To determine the quantity of a radionuclide the disintegration rate must be measured. Since each disintegration is accompanied by a burst of radiation it is necessary only to count these bursts. The number of radiation bursts per second gives a direct indication of the decay rate and thereby the amount of radioactive material present within a given sample.

There are a number of radiation detectors. Among the more common ones are the Geiger tube, the ionization chamber, photographic film, and the scintillation detector. At present the scintillation detector is in universal use in all types of clinical measuring equipment. In fact, it might be said that the whole field of nuclear medicine rests upon the scintillation detector. The commanding position of this instrument is primarily due to its great sensitivity to gamma radiation. The scintillation detector is reliable, fairly rugged, and unfortunately expensive.[1]

In operation the *scintillation detector* converts the energy of incident gamma ray photons into electrical impulses. Each incident photon produces an electrical impulse whose magnitude is proportional to the energy of the given incident photon. The conversion of the energy of a gamma ray photon to the energy of an electrical impulse is a two-step process. The first step takes place in a scintillation crystal. Single crystals of sodium iodide in convenient sizes (2 to 5 inches in diameter) are commonly used as scintillators. A trace amount of thallium is included in the crystal as an activator. The thallium creates imperfections in the crystal structure so that atoms within the crystal may readily assume elevated energy states. When a gamma ray photon strikes such a crystal, energy from the gamma photon is absorbed by the crystal. This absorbed energy causes numbers of atoms within the crystal to assume an excited state in which electrons are moved to higher orbital energy levels. Within a few millionths of a second the excited atoms revert to an unexcited state. The energy associated with the excitation is dissipated as a number of light photons in the visible region of the electromagnetic spectrum. In essence, the *scintillation crystal* is an energy-converting device converting the energy of a gamma ray photon into a flash of light. The intensity of the light flash, or the number of photons in the flash, is proportional to the energy of the gamma ray photon that excited the crystal.

In the second step of the conversion of gamma energy to an electrical impulse, the light flash from the crystal is made to fall upon the cathode of a photomultiplier tube. The *photomultiplier tube* is a high vacuum photoelectric cell that contains a built-in amplifier to increase the electrical current produced by the light incident upon the photo cell. The cathode of the photomultiplier tube is coated with an alkali metal alloy that emits electrons when light photons strike it. The electrons ejected from the cathode strike, in sequence, a series of 10 to 12 charged electrodes in such a manner as to cause additional electrons to be ejected from each electrode. For each electron emitted at the cathode, as many as a thousand electrons may be collected at the final electrode. Note that the height or amplitude of the electrical impulse from the photomultiplier is proportional to the energy of the gamma ray photon that excited the crystal. This electrical impulse may be further amplified by a linear electronic amplifier and ultimately used as a contribution to a numerical determination, or as a point of information in image formation. The conversion of the energy of a gamma ray photon to the energy of an electrical impulse in a scintillation detector is outlined in Figure 1.

Rapid developments are now being made in semiconductor detectors. In time, the semiconductor may challenge the leading position now held by the scintillation detector. *Semiconductor detectors* consist of crystals or combinations of crystalline material whose electrical conductivity is increased from a low to a high value under the influence of radiation. At present the best semiconductor detector for gamma radiation appears to be the lithium-drifted junction detector. A typical *junction detector* consists of two blocks of silicon in contact with each other. One block has been treated with a slight impurity, such as boron, which decreases the number of free electrons in the crystal. The second block has been doped with a minute amount of phosphorus

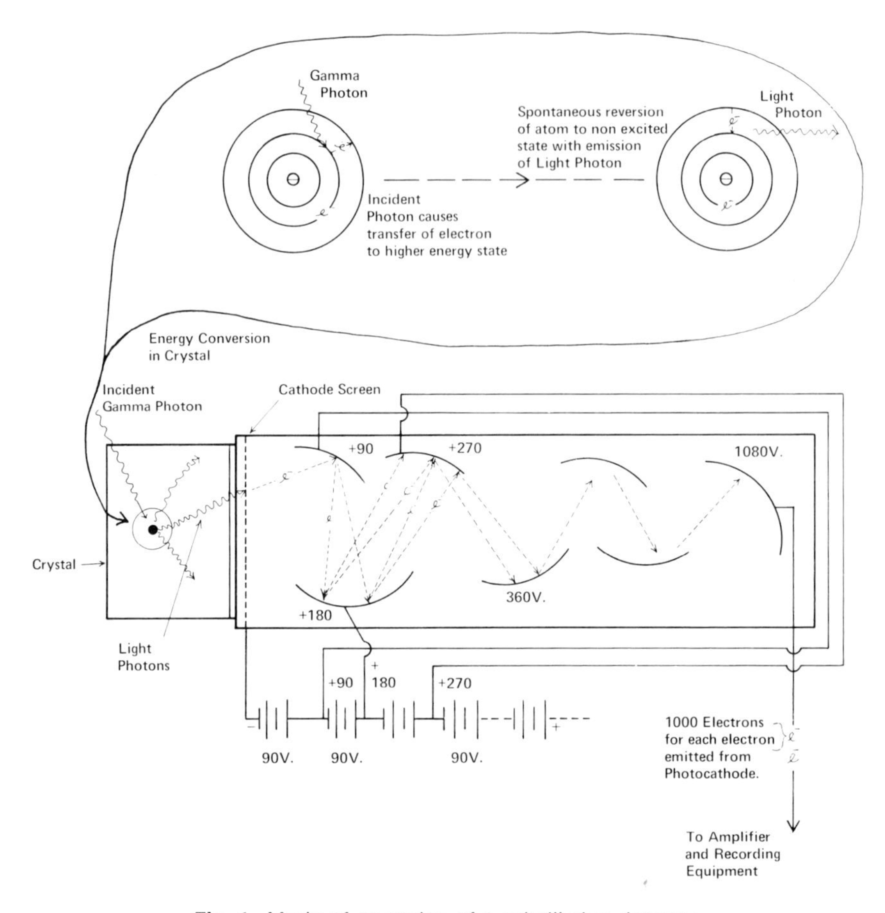

Fig. 1. Mode of operation of a scintillation detector.

or some other impurity, which increases the number of free electrons. When the two crystals are placed in contact under an electrical potential of the correct polarity, little or no current flows. If radiation is now allowed to strike the junction between the two crystals, electrons are released in such a manner that electrical conduction occurs and current flows across the junction. This electrical impulse may be sent to amplifying and recording apparatus in the same manner as electrical impulses from a scintillation detector. The ability of the junction to respond to radiation may be improved by diffusing lithium into the silicon blocks at the surfaces forming the junction. This description of a junction detector is a bit simplified. The reader who desires a greater understanding of semiconductor detectors is advised to consult recent publications on the subject.[2]

IN VITRO TESTING

The measurement of radioactivity in blood, urine, or other samples is usually done with a well counter. A *well counter* is so named because of the scintillation detector design used to sense radiation from within a test tube. The photomultiplier is mounted vertically with the scintillation crystal on top. A hole to receive a test tube is sunk into the crystal. When the test tube is lowered into this hole the lower part of the test tube is surrounded by the crystal, allowing a greater portion of the radiation from the test tube to intercept the crystal. The detector assembly is surrounded by lead to reduce the effect of cosmic rays. Without the shield, cosmic rays and other stray radiation from the environment activate the detector and give counts on the recording system, even when no radioactive sample is near the detector. This unwanted count is called *background.* Background may never be completely eliminated and is always confusing. By shielding the detector with lead, background may be reduced to a point where it is of a minor influence on quantitative measurements.

The electrical impulses from the well detector assembly are amplified and sent to a counting device that keeps a cumulative count of the number of impulses received. A timer is attached to this counter so that counts may be accumulated for a preset time. Thus, if the counter is operated for 1 minute and 2000 counts above background are obtained, then at least 2000 disintegrations must have taken place within the sample during the minute. The number of disintegrations per minute within the sample is directly proportional to the amount of radioactivity present. In many cases the counts arrive so rapidly that the counter can not record them. In such a situation an electronic device is used to scale down the count. Every second or fourth count might be passed to the counter. In some cases only one count in as many as 64 might be recorded. In earlier counters the scaling factor was set by the operator. Modern high speed electronic count indicators have made it possible to work with a single scaling factor that is built into the machine. The combination of scaling device, count register, and clock is often called a *scaler.* This system functions as an integrating device by keeping a record of the total accumulated count.

Pulse Height Analysis

Advantage may be taken of the fact that the amplitude or height of the electrical impulses entering the scaler from the scintillation detector is directly proportional to the energy of the gamma ray photons striking the crystal. Electrical impulses arriving from the photomultiplier and linear amplifier may be passed through a device called a pulse height analyzer before they enter the scaler. The *pulse height analyzer* picks out pulses from photons of particular interest and allows them to enter the scaler for recording, while pulses of the wrong size are rejected. The pulse height analyzer reduces the background count and also makes possible the independent counting of different species of radioactive material present in the same sample.

To understand how the pulse height analyzer selects pulses for counting, consider the following analogy (Fig. 2). Suppose an observer is stationed in the center of a room with a window overlooking a sidewalk. By looking out of the window the observer may see persons passing by. The observer is now instructed to count only those passersby whose heads he can see. If the window is of the correct vertical width and at the proper height in the wall persons 5 to 6 feet in height are counted; those shorter than 5 feet pass by beneath the lower level of the window and are not seen, and those taller than 6 feet are ignored because they extend above the upper level of the window. If the vertical

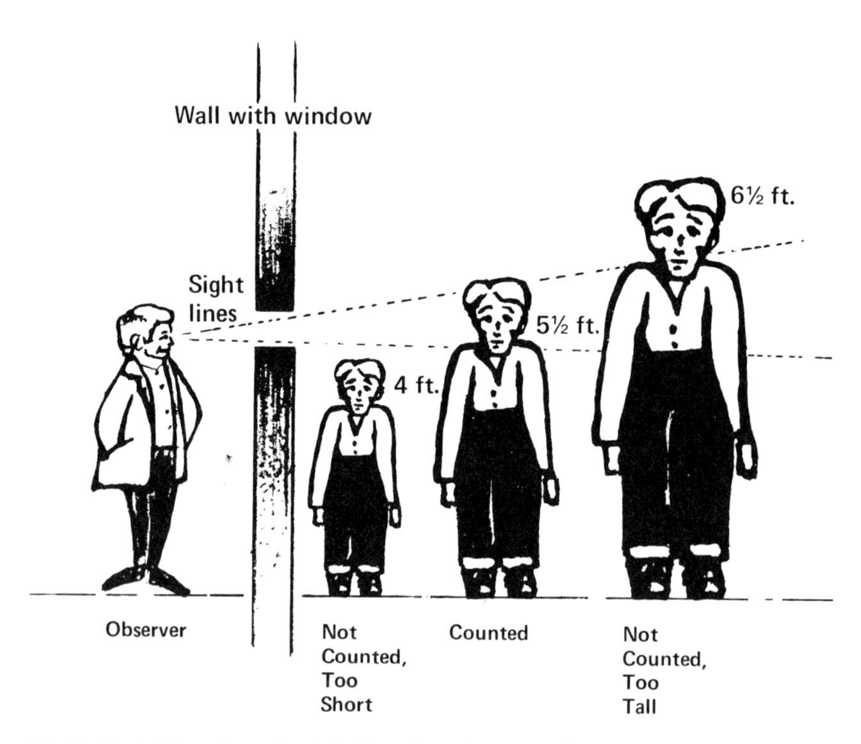

Fig. 2. Basic idea of a pulse height analyzer in terms of an observer looking through a window. Compare with Figure 3.

width of the window is decreased, counting might be limited to persons 5.25 to 5.75 feet tall. If the window were moved up or down the wall a different height group of persons would be selected. The height range of the group selected could be changed by moving the lower window sill only. Selection might also be changed by moving only the upper limit of the window frame. This concept of a window frame is carried over into the designations on the control panel of pulse height analyzers. The pulse height analyzer is seen to be quite like a window in a wall (Fig. 3).

The detector and linear amplifier supply electrical impulses of various amplitudes, or pulse heights, according to the energy of the photons striking the crystal. These impulses are analogous to the people on the sidewalk described above. The electrical impulses are fed to a discriminating amplifier that acts much as the lower frame of a window. This discriminating amplifier responds only if the electrical impulse presented to it exceeds a certain height or amplitude. Note that the discriminating amplifier follows the same type of "all or nothing" law associated with nerve physiology. (In the balance of this discussion the discriminating amplifier is simply called a discriminator.) Sup-

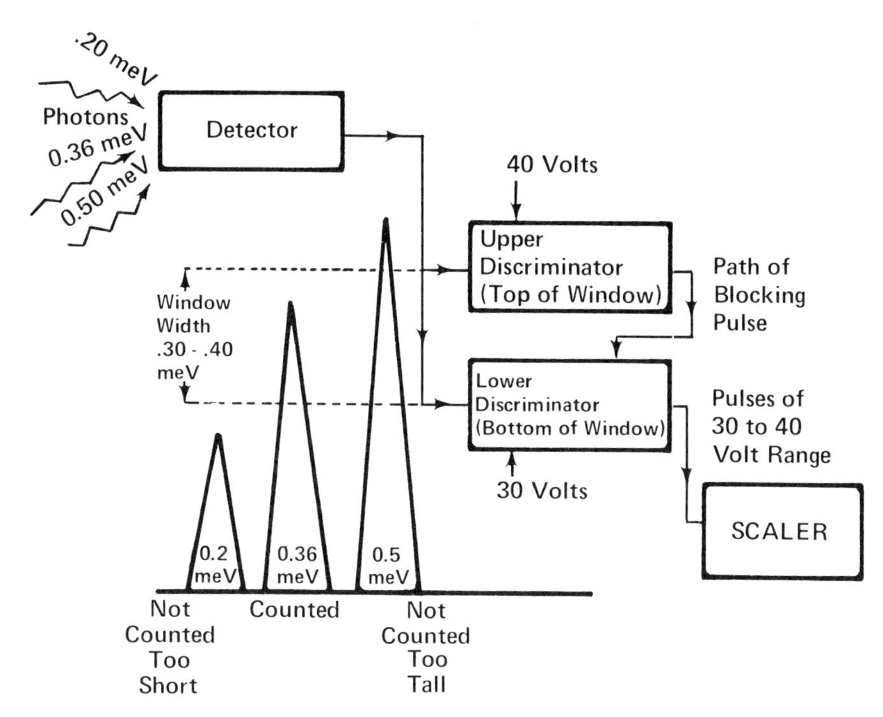

Fig. 3. Components of a pulse height analyzer. Electrical impulses are selected according to amplitude. Compare with Figure 2.

pose we wish to count the 0.364 MeV photons from iodine 131 while excluding most other photons. Assume that the linear amplifier has amplified the iodine pulses to the point where they are at a level of 36.4 volts. In this situation a window covering the range 30 to 40 volts would be practical. The lower level of the window is established by passing the impulses through a discriminator set for 30 volts. Establishment of the upper level of the window is done with a second discriminator set at 40 volts. The action of the upper level discriminator may be followed in Figure 3. Pulses below 30 volts do not trigger either discriminator and do not pass through to the scaler. Pulses of 30 to 40 volts trigger only the lower level discriminator and pass through to the scaler. Pulses above 40 volts trigger both the upper and lower discriminators. When triggered, the upper discriminator sends an electrical impulse into the lower discriminator, which blocks or shuts off the lower discriminator. In this way pulses of over 40 volts are prevented from reaching the scaler. This pair of discriminators form a window that allows passage of pulses 30 to 40 volts in amplitude only. A pair of discriminators acting in this fashion constitute a *pulse height analyzer.* Changing the points at which the discriminators are triggered makes it possible to change window width and position. By changing these parameters, pulses derived from almost any specified photon energy may be sent to the scaler for recording. Pulses other than those within the window are rejected. In some counting equipment separate controls are available, usually designated upper and lower level, for setting pulse acceptance limits. Another arrangement consists of a base line control that moves the whole window up and down by simultaneously raising or lowering both the upper and lower discriminators. A second control (marked "window width") determines the distance between upper and lower levels. In some equipment the discriminators are set to accommodate radiation from a certain nuclide by pressing a button that sets both upper and lower limits. In this case the button is identified with the symbol for the nuclide to be counted (e.g., ^{131}I or ^{198}Au).

The well type detector, pulse height analyzer, and count recording scaler constitute the usual apparatus for measuring the radioactivity in gamma-emitting test tube samples (Fig. 4). The pulse height analyzer is set for the energy level of the nuclide to be measured. The counter is set to zero. The required counting time is set, and the sample to be measured is placed in the well. The timer is now started and allowed to run until it stops. The accumulated count for the running period is indicated when the counter stops. If the number of counts per minute is needed the accumulated count is divided by the number of minutes in the counting period. The activity of an unknown is usually determined by comparing it with a known sample of the same material. The standard should have about the same activity as the unknown. Both the unknown and the standard are corrected for the presence of background count by subtracting that count.

Suppose a 2-ml unknown sample of ^{131}I gave a count of 5,050 in 1 minute

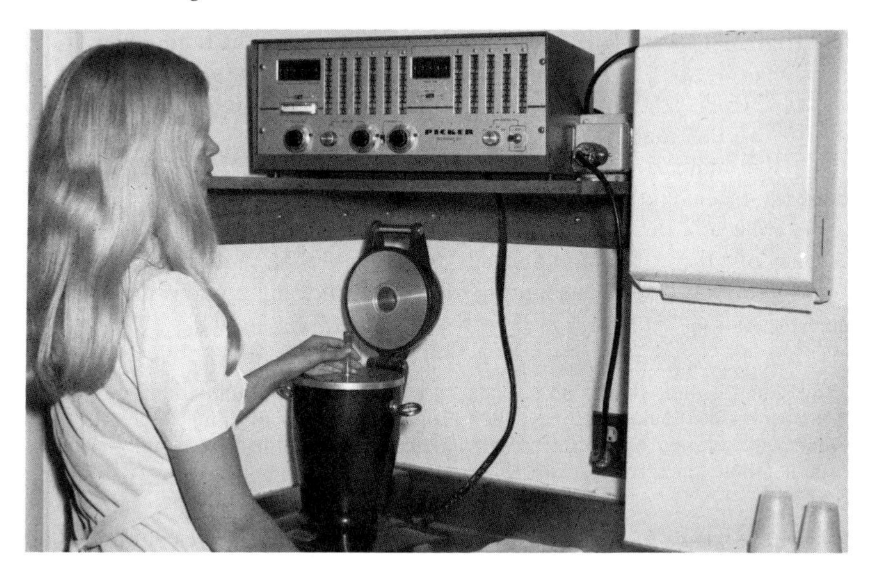

Fig. 4. Well counter with pulse height analyzer scaler and timer. (Courtesy of Rochester General Hospital, Rochester, NY)

and a 2-ml standard of 0.5 μCi gave 10,050 counts in 1 minute. A background count of 50/minute was obtained with nothing in the well. The unknown activity is then:

$$\frac{\text{Unknown count} - \text{background}}{\text{Standard count} - \text{background}} \times \text{amount of standard}$$

$$\text{Unknown} = \frac{5050 - 50}{10{,}050 - 50} \times 0.5\ \mu\text{Ci} = 0.25\ \mu\text{Ci}$$

Background should always be subtracted from any count before division or any other mathematical operation is performed. The unknown and the standard should be in identical test tubes and have the same volume. In general terms the geometry of the standard and unknown should be the same. Identical geometry ensures identical radiation absorption and scatter for the two samples.

Statistics of Counting

One soon notes a disconcerting fact. If repeated measurements are made on the same radioactive sample, the counts are not identical. A sample might give 100 counts on the first trial, 112 on the second, 97 on the third trial, and so on. This fluctuation in count is due to the random nature of radioactive decay.[3] The disintegration of radioactive atoms is a matter of chance. There is a 50-50 chance that a given atom will disintegrate during one half-life of the material. This chance-decay law is characteristic of all radioactive material. The count obtained in a given period of time varies about the true count rate in a statistical manner. The greater the number of counts collected, the closer is the indicated count to the true count. Thus the accuracy of a recorded count depends only upon the number of counts recorded and in no way upon the counting time. The following ground rules prevent one from falling into grievous error due to counting statistics:

1. The sample count should be at least 10 times the background count.
2. The net count should be at least 1000 counts to have any meaning. A count of less than 1,000 should be considered an indeterminate reading (net count = sample count minus background count).

On the basis of a 99 percent confidence level, a net sample count of 1,000 corresponds to a statistical error of 10 percent in the result. Below 1,000 counts the error becomes progressively greater. A net sample count of 5,000 corresponds to a statistical error of ± 5 percent. An accuracy of ± 3 percent is obtained with 10,000 counts. For most samples 5,000 to 10,000 counts are needed for acceptable results. If the analysis of error, the design of experiments, or the testing of counting equipment is contemplated, additional knowledge of the statistics of probability processes is needed.[4]

So far only the counting of gamma-emitting samples has been considered. For samples that emit beta radiation only, particularly that of low energy, a different type of detection system is used in a well counter. This system for beta counting is referred to as a liquid scintillation counter.[5] In the *liquid scintillation counter* the scintillation crystal is eliminated and the well is modified so that a vial may be placed directly in front of the photomultiplier tube in the position of the crystal. The beta-emitting sample in liquid form is mixed with a liquid scintillating material. This combination of sample and scintillator emits flashes of light as disintegrations take place in the sample. The flashes of light from the sample vial act upon the photomultiplier tube in the same manner as light from the iodide crystal. Since the sample and scintillation material are in intimate contact there is a high efficiency of energy transfer from source to scintillator. There are a number of organic materials that may be used as the scintillating agent. A common one is 1,5-diphenyloxazole in xylene. Even though special sample preparation is provided, the light flashes from all of the known liquid scintillators are very weak. In fact, two photomultiplier tubes in a special coincidence counting circuit are frequently used to take full advantage of the minute flashes of light from the liquid scintillator. The liquid scintillation counter is most often found in research laboratories where weak beta emitters such as hydrogen 3 and carbon 14 are in use.

IN VIVO MEASUREMENTS

For quantitative measurements of radioactivity within the body a collimated detector is substituted for the well detector. The same pulse height analyzer and scaler used for the well may be used with the collimated scintillation detector. The *collimator* is a device made from radiation-absorbing material, such as lead, which defines the dimensions and direction of a beam of radiation. When measuring with a radiation detector the purpose of the collimator is to limit detected photons to those which come from a particular location within the patient. The collimated system consists of a crystal and photomultiplier housed in lead, as in the well counter. In the collimated detector the front end of the crystal is left open and is attached to a lead cone. Radiation entering the conical collimator falls upon the crystal and is detected. Radiation from directions other than the open end of the cone is absorbed in the lead of the cone or in the lead housing around the detector. The collimating cone may have various dimensions; its size and shape should be such as to provide a region of almost equal detector response to all points of the area covered by the cone. Such a cone is called a *flat-field collimator.*

A typical measurement of internal radioactivity is the thyroid uptake. The usual procedure is to count with a standardized flat-field collimator over the patient's neck, the source of the unknown. A count with the same detector and collimator is then made over a source of known radioactivity located in a standard thyroid uptake neck phantom. A collimator with a projected 15 cm diameter flat field at the thyroid level when the distance between the detector and patient is 25 cm is standard for thyroid uptake. The standard thyroid uptake *neck phantom* is a solid Lucite cylinder 15 cm in diameter by 15 cm in height.[6] A cavity 51 mm in diameter by 102 mm deep is machined into the end of the cylinder 0.5 cm from the front surface to accomodate a 30-ml bottle. The known source (an amount of ^{131}I identical to that given to the patient) is diluted to 30 ml and placed in the neck phantom. The neck phantom provides a geometry around the standard equivalent to an average thyroid in an average neck. A Lucite capsule holder and a known ^{131}I capsule may be substituted for the bottle with acceptable deviation from specified geometry. The object is to keep the distance, source dimensions, and scatter material equivalent so that radiation absorption and scatter around the standard and unknown are nearly identical.

In performing a 24-hour thyroid uptake the following procedures and calculations are carried out. The patient is given a capsule of sodium iodide 131, and an identical capsule is set aside as a standard. Twenty-four hours later a count is made over the patient's neck; a second count is made over the standard in the phantom neck; and a third count with nothing in front of the detector is made to determine background. The percent of iodine uptake by the thyroid is computed as follows:

$$\frac{\text{Patient's count} - \text{background}}{\text{Standard count} - \text{background}} \times 100 = \%\ \text{uptake}$$

Note that no correction for radioactive decay is needed because the iodine in the standard has been decaying at a rate identical to the iodine in the patient. The two losses cancel each other out when a ratio is taken between the unknown and standard. Once again attention is called to the fact that background should be subtracted and net count obtained before multiplication, division, etc.

DYNAMIC FUNCTION STUDIES

So far we have considered the measurement of accumulated quantities of radioactive material that are more or less static in nature. In some circumstances it may be desirable to determine the rate of accumulation or loss of radioactive material within a patient. Measurements of this nature may be made with a flat-field collimator and detector of the type used for thyroid uptake. The same amplifier and pulse height analyzer used for thyroid uptake may be used for dynamic function studies. Another device, the ratemeter, must be substituted for the scaler and count register. Remember that the scaling counter keeps a record of the total accumulated count without regard to the count rate. The *ratemeter,* unlike the counter, indicates only the number of counts per minute being received at any particular moment.

The Ratemeter

To understand the properties of the ratemeter and its relationship to the scaling count register, consider a widely known ratemeter—the automobile speedometer. The speedometer tells a driver the rate of speed of his car at any moment in terms of miles per hour. By watching the speedometer the maximum and minimum rates of speed on a trip may be determined. The average speed may be estimated and the distance traveled computed by multiplying the average speed by the time for the trip. Distance determined in this way is subject to large error due to the estimation of average speed. For example, in an automobile traveling in 30 mile an hour traffic for 1 hour stops may be made at traffic lights and the speed varies as the car passes another or is forced to slow at congested points. In attempting to determine distance traveled by watching the speedometer, the error might easily be as much as 2 or 3 miles. On the other hand the maximum and minimum rates of speed or the rate of speed at any given time may be accurately determined by watching the speedometer. Along with the speedometer is usually an odometer that indicates how far the car has traveled. If at the start of the hour of travel the odometer reading was 3,000 miles and at the end of the hour the reading was 3,030 then the distance traveled is quite accurately known to be 30 miles. The odometer tells nothing about the rate of speed along the way. The car could have traveled at 30 miles an hour for an hour; or it could have traveled at 60 miles an hour for 15 minutes, stopped for half an hour, then continued for 15 minutes at 60 miles an hour. The odometer reading would be 30 miles at the end of an hour in either case. The scaling count register is like the odometer. Counts obtained upon the counter are analogous to distance traveled. The ratemeter used to determine radioactive count rate, like the speedometer, indicates the count rate at any particular moment. In test tube measurements or tests like thyroid uptake total count is determined more accurately with the counter than with a ratemeter. At times it is desirable to know the rate at which isotope location or concentration is changing. Here the ratemeter is needed and the counter is of little or no assistance.

There are two types of ratemeters. The digital ratemeter is actually a scaling counter designed to count for a short period of time, record the results, and repeat. A digital ratemeter may count for fractions of a second to a few seconds. The digital ratemeter has certain advantages of a statistical nature, but its high cost has kept it from being widely used.

The analog ratemeter is less expensive and is adequate for most clinical studies. It gives a continuous rather than an intermittent reading as does the digital meter. The analog ratemeter receives electrical impulses at random from a pulse height analyzer or detector. These random electrical impulses are fed into an electrical circuit that responds to the average frequency of incoming pulses by giving a continuous electrical current proportional to the average count rate. This current is then passed through the rate-indicating meter. The electrical circuit of the analog ratemeter contains a capacitor, and each count adds a given amount of electrical charge to this device. A resistor is placed across the capacitor so that electrical charge leaves the capacitor in a slow and continuous fashion. The count rate is indicated by the amount of charge in the capacitor at any moment. The resistor-capacitor combination acts as a temporary memory whose characteristics may be expressed in terms of a time constant.

By analogy the capacitor behaves as a bucket, the resistor as a hole in the bottom, and the counts as cups of water thrown into the bucket. As each cup of water (count) is thrown into the bucket the water level rises. If the input of water is halted the water soon drains out leaving the bucket empty. (The count rate corresponds to the water level in the bucket.) The faster the water is thrown in, the higher is the level in the bucket. (A high level of water in the bucket corresponds to a high count rate on a ratemeter.) The response of this leaky bucket may be made slower, or of a longer time constant, by making the bucket large and the hole small; or it may be made shorter by using a small bucket or larger hole. If this bucket analogy is extended to the scaling counter it is seen that the counter acts as a bucket without a hole: As the water level continues to build so long as water is thrown in, so the indicated count continues to build as long as counts are fed to the counter.

By changing the resistance or capacitance in the ratemeter circuit, the time constant may be made shorter or longer with the following results:

1. If the time constant is increased:
   a. The meter becomes more sluggish and is less able to follow rapid changes.
   b. The reliability of the reading becomes statistically greater.
2. If the time constant is decreased:
   a. The meter becomes better able to follow rapid changes.
   b. The statistical reliability of the readings is decreased.

The maximum accuracy of a ratemeter in a given situation is obtained through an optimum compromise between ability to respond rapidly and the statistical reliability of the reading. Important characteristics of ratemeters are as follows:

1. The count rate is presented as an average of the counts contained in a temporary memory. This memory is considered temporary because counts are continuously removed from it at a rate proportional to the frequency with which counts are added.
2. Statistical accuracy is proportional to the number of counts in the memory at any time; the accuracy may be calculated from the count rate and the time constant.
3. The correct response, for practical purposes, is obtained in a time equal to five times the time constant. Five time constants yield 99.4 percent of the true reading.

A comparison of the count-accumulating scaler and analog ratemeter is best carried out by example. Consider the response of a digital scaler and a ratemeter to a signal of 100 counts per second (cps). Assume the ratemeter has a time constant of 1 second. Starting with zero radiation both instruments are exposed to a source of radiation giving 100 cps to each instrument. At the end of 1 second the radiation ceases. The response of the instruments is shown in Figure 5. After 1 second, as the radiation ceases, the scaler reads 100 counts. At the end of the identical 1 second the ratemeter reads 63.3 cps. At the end of 2 seconds (1 second after exposure has ceased) the scaler continues to read 100 counts. During this time the ratemeter reading drops to 23.2 cps. If the ratemeter had a time constant of 0.5 seconds, instead of 1.0 second, the reading at the end of the first second would be 86.5 cps and at the end of 2 seconds 11.7 cps.

If a continuous source of radiation is presented to a scaler the indicated count increases continuously until the radiation is removed or the scaler halted. A ratemeter responds in the following manner if exposed to a continuous source of radiation. Starting from zero the reading increases in the following way:

One time constant yields 63.3 percent of final reading.
Two time constants yield 86.5 percent of final reading.
Three time constants yield 95.0 percent of final reading.
Four time constants yield 98.2 percent of final reading.
Five time constants yield 99.4 percent of final reading.

When the final reading is reached it continues to be indicated until the radiation is removed, at which time the reading of the ratemeter decreases in an inverse manner to the buildup of the reading upon exposure to radiation. In a

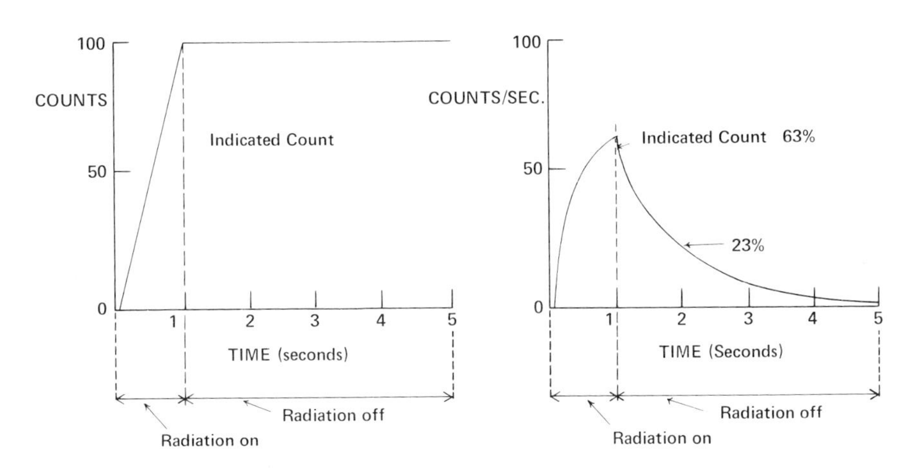

Fig. 5. Response of a count-accumulating scaler (left) and that of an analog ratemeter (right) to a 1-second exposure to radiation.

time equal to one time constant after radiation has ceased, a ratemeter reads 36.7 percent of its previous maximum indication. In a time equal to five time constants after the radiation has stopped, the meter again essentially reads zero (0.6 percent of maximum).

Two controls not found on a scaling counter are usually associated with an analog ratemeter. The first is a range switch, which changes the response of the meter so that the counts per minute received fall within a range covered by the meter. For example, a given meter scale reads from 0 to 300 counts per minute (cpm). The range switch might provide a multiplying factor for the scale of × 1, × 10, or × 100, which would provide count ranges of 0 to 300, 0 to 3,000, or 0 to 30,000 cpm, respectively. The second control is the time constant setting. This may be made to vary from a fraction of a second to 100 seconds or more. The range available on any particular meter depends upon the use for which it was intended. Certain guidelines must be followed to obtain acceptable results when using a ratemeter with an adjustable time constant. The guidelines are as follows:

1. When a ratemeter is used to determine the activity of a nonvarying source, after the source is positioned wait three times the time constant before reading the meter. Where maximum accuracy is required, wait five times the time constant before reading.
2. When a ratemeter is used to record transient phenomena the time constant should be less than one-third the shortest expected transient time for acceptable accuracy. For maximum accuracy the time constant should be less than one-fifth the transient time.
3. To ensure reasonable statistical accuracy the memory of the ratemeter should contain over 1,000 counts at the time of reading.
   a. Counts in memory = 2 × time constant (seconds) × count rate (cps)
   b. When a satisfactory compromise between statistical accuracy and response time cannot be achieved with a given dose of radioactivity, the dose must be increased.

The renogram is typical of a study where the ratemeter is used (Chap. 4). An identical counting and recording chain is set up for each kidney. Each counting channel consists of a detector with a flat-field collimator followed by an amplifier and pulse height analyzer that feed a recording ratemeter. A *recording ratemeter* is one in which the rate-indicating needle is fixed to a pen that traces an indication of the count rate on moving graph paper, thereby providing a permanent record of the count rate at any time during the test period. To perform the renogram a scintillation detector probe is positioned over each kidney. Radioiodine-labeled orthoiodohippurate is injected intravenously. The graphic rate recorders are started at the time of injection. The recorders follow the transit of the ^{131}I-orthoiodohippurate through the kidneys as reflected by gamma radiation from the iodine label. The duration and characteristics of the components of the renogram tracing are evident from watching the rate recorders, and a permanent record is obtained upon the graphic chart. A comparison of the graphs (for both kidneys) with a normal renogram provides diagnostic information.

In a typical renogram a peak count rate of 10,000 cpm might be obtained. If, for example, a time constant of 4 seconds is used the statistical error in determining the peak value may be calculated as follows:

$$\frac{10{,}000 \text{ cpm}}{60 \text{ seconds}} = 166.6 \text{ cps}$$

Counts in memory = 2 X 4-second time constant X 166.6 cps = 1333 counts

The statistical error for 1,333 counts in memory has a high probability of being within ± 3 percent and is almost certain to be within ± 8 percent.

For practical purposes the statistical reliability of ratemeter readings may be considered certain to fall within the following ranges.

Counts in memory (No.)	*Statistical deviation (%)*
100	± 30
1,000	± 9.5
5,000	± 4.2
10,000	± 3.0

REFERENCES

1. Cradduck TD: Fundamentals of scintillation counting. Semin Nucl Med 3:205, 1973
2. Hoffer PB, Beck RN, Gottschalk A (eds): Semiconductor Detectors in the Future of Nuclear Medicine. New York, Society of Nuclear Medicine, 1973
3. Chase GD, Rabinowitz JL: Principles of Radioisotope Methodology, 3rd ed. Minneapolis, Burgess, 1967, pp 75-108
4. Bahn AK: Basic Medical Statistics. New York, Grune & Stratton, 1972
5. Wang CH, Willis DL: Radiotracer Methodology in Biological Sciences. Englewood Cliffs, NJ, Prentice-Hall, 1965, pp 104-143
6. National Bureau of Standards Handbook 86: Radioactivity (ICRU) Report 10 c. Washington, DC, Government Printing Office, 1963, pp 29-31

chapter 10

SCANNING AND IMAGING EQUIPMENT

Important diagnostic information may be obtained by visualizing the distribution pattern of a radionuclide in a field of interest, such as the thyroid. It would seem a simple matter to set up a camera, focus the lens upon the area, and let the camera record an image of the radioactivity in that area. Such an attempt, however, fails for two reasons. First, the gamma radiation coming from the patient travels in straight lines in all media and can not be focused by a lens. Second, the intensity of the radiation from the patient is of such a low level that the formation of an image on the most sensitive film would require an impossibly long time.

To circumvent these limitations it is necessary to scan the area in such a manner as to obtain information from which an image may be constructed. To do this, a point by point dissection of the pattern of radiation coming from the patient is made. Two types of information from each isolated point in the region are needed for image formation. First, the location of the point must be determined so that it can be correctly placed in the image. Second, the intensity of the radiation from that point must be determined so that a proportionate density is recorded in the image. Both moving and stationary radiation-detecting devices may be used to collect information for image construction.

The process of collecting image-forming information from the spatial distribution pattern of a radionuclide in the field of interest, such as the thyroid. is called *scanning*. The device that collects the image-forming information is referred to as a *scanner*.

MOVING DETECTOR SYSTEMS

The moving detector is the basis of the rectilinear scanner, which is in wide use. The mechanical arrangement of the rectilinear scanner is seen in Figure 1 and the scanner itself in Figure 2. A bar with a scintillation detector and collimator on one end is positioned so that radiation emanating from a point within a patient passes through the collimator into the detector. Electrical impulses arising in the detector are amplified and passed to the image-printing devices on the opposite end of the bar. In operation the bar is set into a back and forth motion, the detector traveling transversely across the area of interest. At the end of the transverse motion a longitudinal shift is made and the detector then travels transversely in a direction opposite to the original motion (Fig. 1). This cycle is repeated until the desired area has been scanned by the detector. The motion of the detector thus performs a point by point dissection of the radiation pattern emanating from the patient.

Information pertaining to the location of various points within the

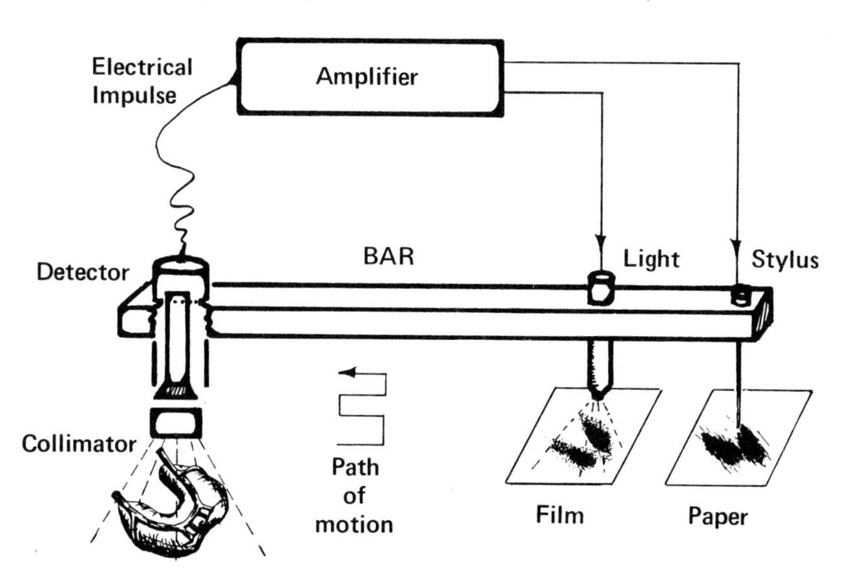

Fig. 1. Mode of operation of a rectilinear scanner.

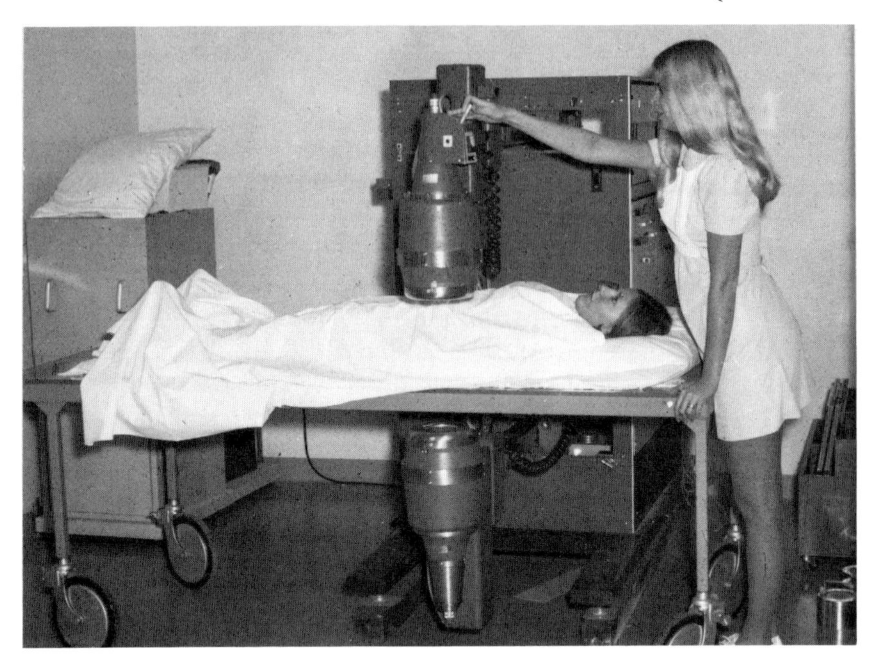

Fig. 2. Rectilinear scanner equipped with two scanning heads and two independent printing systems. Anterior and posterior views can be obtained simultaneously. (Courtesy of Rochester General Hospital, Rochester, NY)

patient is transferred to the printing devices from the detector by the motion of the bar, which is common to both the detector and printers. Information regarding the intensity of radiation from each given point is transferred by the electrical impulses from the detector to the printers. The printing devices pass over paper or photographic film as the detector passes over the patient. In so doing, the printers construct an image from information abstracted from the patient by the detector as it moves over the various points. During image formation radiation intensity information is passed to the printing equipment in the form of electrical impulses which correspond to radiation photons emitted by the radioactive material within the patient. Each electrical impulse is considered as one count. Intensity information is usually received at the rate of several thousand counts per minute. The detector is typically passed over the patient at a speed that provides about 800 counts per square centimeter in the areas of greatest radioactivity.

Successful operation of the rectilinear scanner depends upon certain special characteristics provided by the various components of the scanning system. The more important of these characteristics are considered here, starting with the function of the collimator.[1]

The collimator on a rectilinear scanner usually consists of a cylindrical lead block, 3 to 5 inches in diameter, through which hundreds of holes have been bored. The holes are located at an angle on a specified radius of curvature so that they all converge on a given point 3 to 5 inches in front of the detector (Fig. 3).

Although such a collimator does not focus, it does tend to accept radiation from points at a given distance in front of it; hence it is often called a *focusing collimator*. Much of the radiation from a point closer than a given distance is absorbed by the collimator (Fig. 4). At the proper distance, radiation from a point in front of the collimator passes through all of the holes. Beyond this so-called focal distance, radiation again is absorbed except for the central hole (Fig. 5). Thus radiation emitted from a horizontal plane within the patient may be selected for image formation. A scan performed on a rectilinear scanner has more or less the properties of a body section radiograph because of the selective action of the collimator.

Eliminating background counts contributes to the clarity of the image, so a pulse height analyzer is included in the amplification system. Even so,

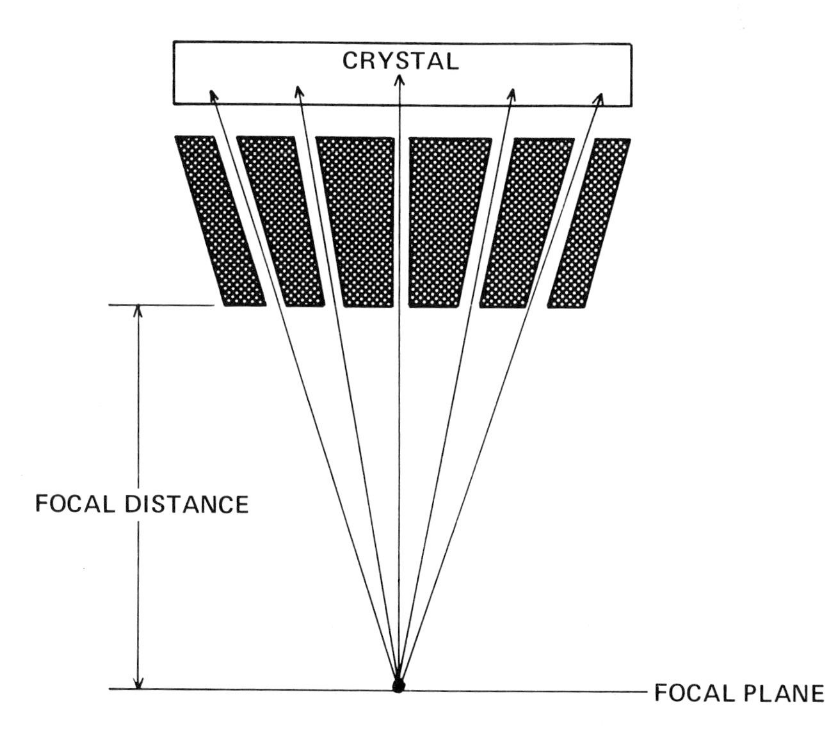

Fig. 3. Pathway of gamma ray photons through a collimator with the source at the focal distance. Compare with Figures 4 and 5.

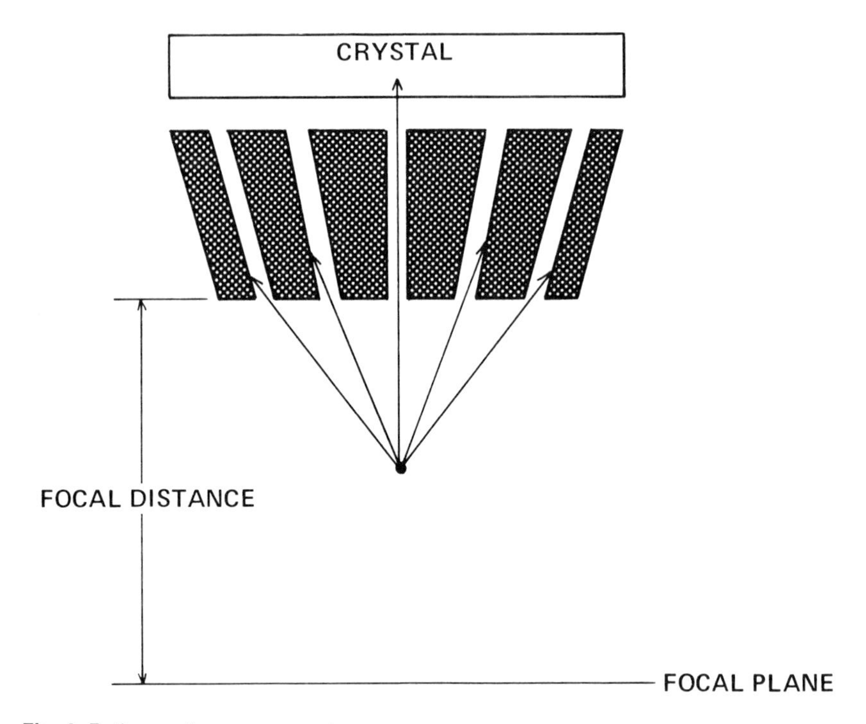

Fig. 4. Pathway of gamma ray photons through a collimator with the source closer than the focal distance. Compare with Figures 3 and 5.

background is often annoying and a second device variously referred to as a background suppressor, background eraser, or background cutoff is used. This device is capable of responding to changes in the count rate; when the count rate is low the instrument reacts by turning the printing devices off, and when the rate exceeds a certain value the printing devices are turned on by the *background erase circuit.* By using background erasure the scanned organ stands out more clearly in the image, not being surrounded by random (background) dots.

The counts received by the recording system form dots on paper or film. Those on paper are made by a mechanism similar to that of a typewriter; dots of uniform blackness are made as counts are recorded, and the image is interpreted in terms of numbers of dots present in various areas of the image. In the case of the film image a lamp is flashed exposing a small area of the film, thereby making a dot. Dots of varying degrees of blackness are made by controlling the brilliance of the exposure lamp. In practice the brilliance of the lamp is controlled by an electronic circuit which follows the count rate. When the count rate is high the dots are made darker, and when lower the

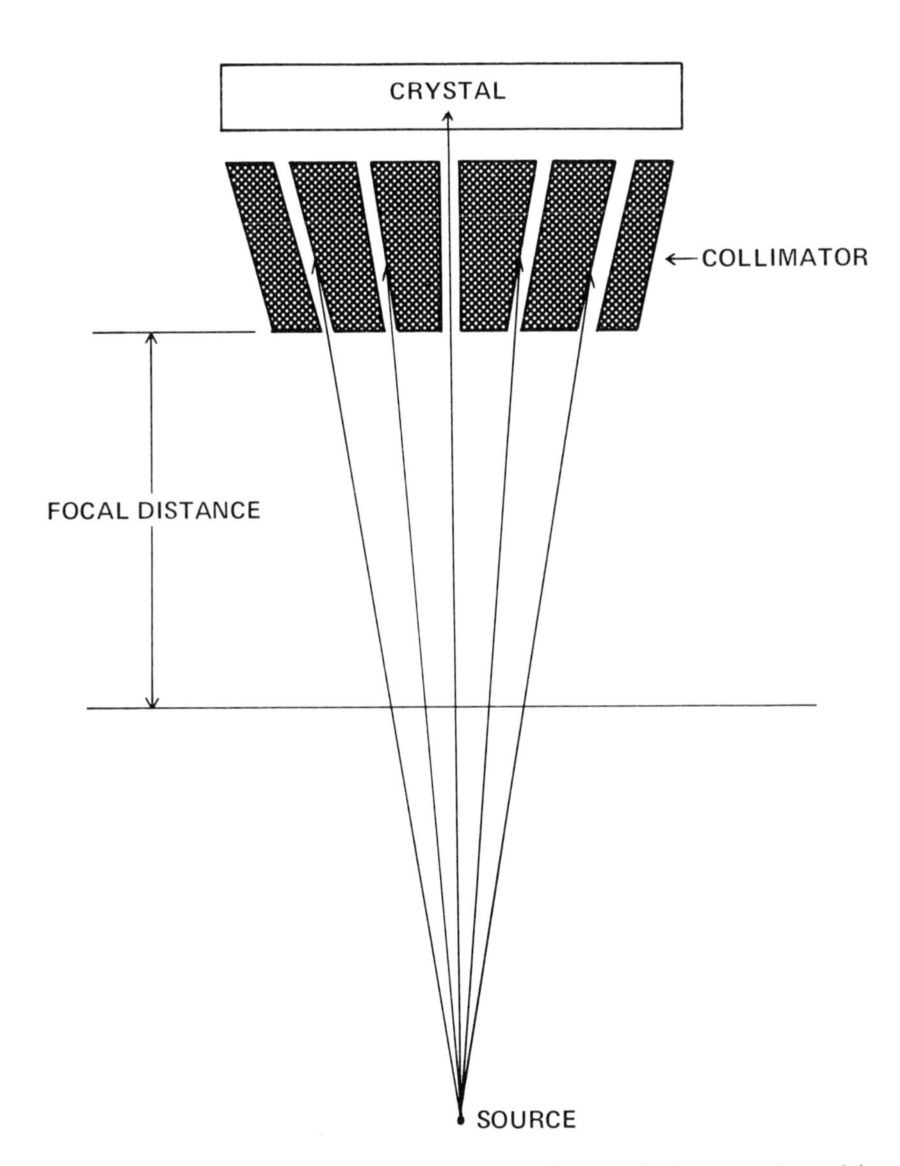

Fig. 5. Pathway of gamma ray photons through a collimator with the source beyond the focal distance. Compare with Figures 3 and 4.

dots are lighter. Thus the number of dots and their darkness must both be evaluated in interpreting film scans.

An optional device available on some scanners provides a color print. With a color printer various numbers of dots are made, as in paper or photo printers, but there is an additional mechanism that changes the color of the dots according to the count rate. Thus when counts are received at a high rate the dots are red; at a slightly lesser rate they are orange; and as the rate decreases the color changes through yellow, green, blue, and purple.

STATIONARY IMAGING DEVICES

There are several types of stationary imaging devices in which neither the patient nor the detector is moved during the image-forming period. In these systems both radiation location and intensity information from various points within the body must be provided by the detector. The gamma camera (Fig. 6) is the stationary device in widest use at present, so it is used here to explain the principles of stationary imaging equipment.[2]

Image dissection in the gamma camera is performed by the collimator, which is positioned just above the patient at the area of interest. The collimator in the gamma camera is a lead plate 10 to 12 inches in diameter and of a thickness sufficient to absorb gamma radiation from the radionuclide within the patient. The lead plate has thousands of parallel holes so that it appears much like a honeycomb. Radiation parallel to the holes passes through them, while radiation striking the collimator at an angle is absorbed by the septa between the holes. Above the collimator is a scintillation crystal equal in diameter to the collimator and about half an inch thick. As radiation photons pass through the collimator and strike the crystal flashes of light are produced within the crystal at points corresponding to collimator holes. An array of some 19 photomultiplier tubes is positioned above the crystal to detect light flashes within the crystal. The mounting of the photomultiplier tubes is so designed that all tubes within the array are more or less affected by a flash of light occurring any place within the crystal. The tube directly above the flash receives the most light and gives the largest electrical impulse. Adjacent tubes recieve less light according to their relative locations. Correspondingly, smaller electrical impulses are delivered by the adjacent tubes. The electrical impulses from all 19 photomultipliers are passed through a resistance network in such a manner that four derived electrical impulses are obtained. These impulses carry location information for a particular radiation event. Two electrical impulses specify transverse position, and the other two longitudinal position. A fifth electrical impulse is derived by summation from all the photomultipliers. This fifth impulse corresponds

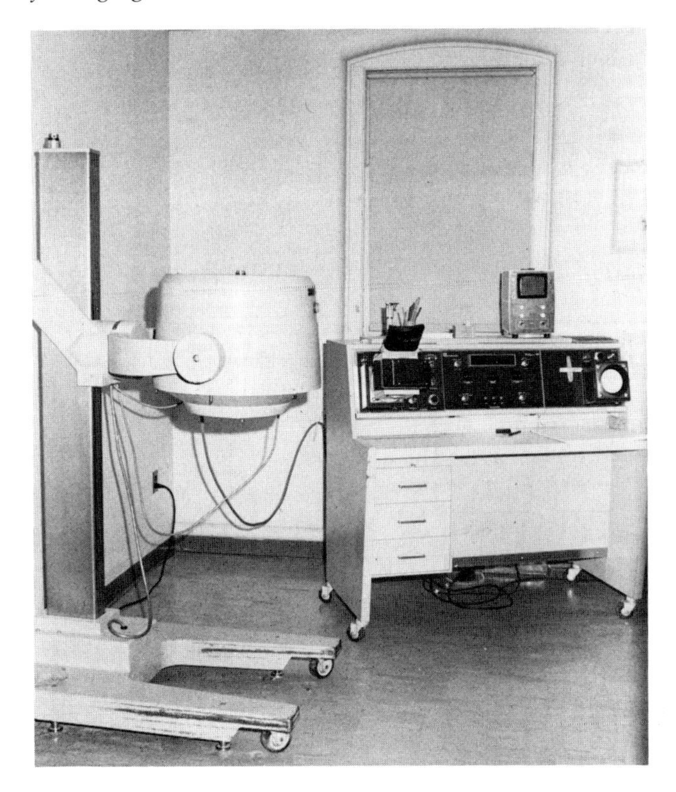

Fig. 6. Typical gamma camera installation. (Courtesy of Highland Hospital, Rochester, NY)

to the total amount of light received by all tubes and carries intensity information. All five of the pulses are passed through computer circuits that transform the pulses into a form suitable for operation of a cathode ray oscilloscope. The oscilloscope is equipped with a camera that provides image printout.

The operation of the gamma camera may be clarified by following the sequence of events in Figure 7. First, radioactive disintegrations produce photons at a point within the patient. Although these photons travel outward in all directions, only those going upward toward the scintillation crystal are involved in image formation. The collimator prevents photons from entering the crystal except at a point directly above the location of photon origin. In this manner information on the location of photon origin is passed to the crystal. A flash of light occurs at a point in the crystal corresponding to the specified point within the patient. The flash of light in the crystal is too weak to be seen or recorded on photographic film, but

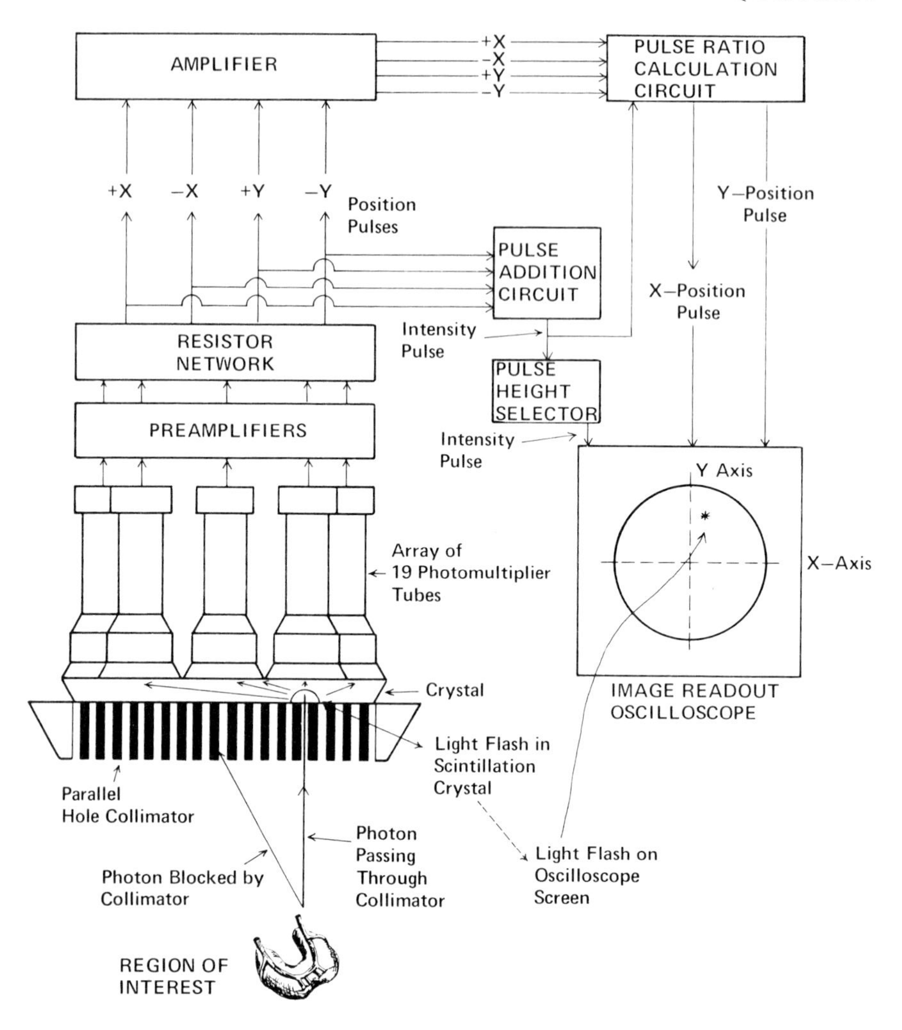

Fig. 7. Mode of operation of the components of the gamma camera.

photomultiplier tubes sense it and respond with electrical impulses. The tubes near the flash respond with large impulses, and those more remote with smaller impulses. This relative response of the photomultipliers indicates the location of the light flash in the crystal. The electrical impulses from the photomultiplier tubes are amplified. The flash location information they contain is converted into electrical currents suitable for deflecting the electron beam in a cathode ray oscilloscope. By this means the electron beam in an image readout oscilloscope is directed to a point on the screen that corresponds to the flash location in the cyrstal. At this point a flash of high light intensity is produced on the oscilloscope screen by the pulse which carries intensity information. If a camera is located in front of the oscilloscope screen the flash of light is recorded on film.

This entire process is repeated hundreds to thousands of times each second with radiation photons originating at multiple points throughout the region of interest within the patient. In this fashion a picture of the organ in question is built up on the film in the camera focused on the oscilloscope screen. The final image is composed of 1,000 to 100,000 dots that correspond to light flashes emanating from the scintillation crystal.

A second stationary device is the autofluoroscope.[3] This instrument varies from the gamma camera in that it makes use of a number of small crystals, rather than one large one, as the radiation-detecting element. The autofluoroscope employs an array of 260 sodium iodide crystals 3/8 inch square by 2 inches in length. This crystal array is packed into an area 6 by 9 inches. The centers of the crystals are about 1 cm apart and correspond to appropriate holes in the collimator. Light flashes from the crystals are directed over plastic light guides to a bank of 33 photomultiplier tubes. Light from each crystal is split into two parts in the light guide system. The two portions of light are directed to two photomultiplier tubes which designate event location. Electrical impulses from pairs of tubes, responding to crystal flashes, are sent to a magnetic core memory where radiation event location is recorded by a rank and file system. Intensity information is also collected and recorded. After all of the information pertaining to an area is recorded it is fed to an oscilloscope for visualization. The image on the oscilloscope can be observed or photographed immediately, or the contents of the magnetic memory can be stored on magnetic tape for future recall.

A third type of imaging system is built around an image intensifier tube of the type used in x-ray fluoroscopy.[4] Since this system is not widely used it is not described here.

Comparison of Imaging Systems

A comparison of moving and stationary imaging systems reveals certain advantages and limitations inherent to each. The rectilinear scanner has a low radiation detection efficiency because radiation from only one point in the patient is seen by the detector. Most of the radiation from the patient fails to enter the detector and is wasted. The detector must slowly move back and forth to explore a given area. The gamma camera, on the other hand, receives radiation simultaneously from the entire area under observation, so much more of the emitted radiation is utilized. A rectilinear scanner requires several minutes to half an hour to produce an image, while the gamma camera yields an image within seconds to a few minutes. The speed of image formation is a definite advantage of the gamma camera. Image formation is fast enough to permit dynamic function studies through serial views of an organ, while the rectilinear scanner is limited to views of static situations. The speed of a gamma camera is also advantageous in busy departments with a high volume of work.

An important aspect of scanner performance is resolving power. *Resolving power* is a measure of the ability of a scanner to record detail. Resolution is often measured with line sources of radiation, which are made by filling capillary hematocrit tubes with radioactive solutions. Two line sources are placed in a parallel position in front of a scanner with a horizontal distance of 1 to 2 cm between them. Two lines of activity are seen on a scan. If scans are repeated as the two sources are moved closer together, a point is reached where the two lines in the image merge, and the scanner is no longer able to separate the two sources. The scanner's resolution is designated as the distance between the sources when they are last seen as two entities in the image. A resolution of 5 to 7 mm between lines characterizes a good performance for either a rectilinear scanner or a gamma camera.

In a rectilinear scanner the region of best resolution is at the plane of collimator focus, which is usually 3 to 5 inches beyond the front surface of the collimator. At nearer or greater distances both resolution and sensitivity decrease. The image produced by a rectilinear scanner has some of the properties of a tomogram or body section radiograph. An area of increased or decreased radioactivity is seen with maximum clarity if it lies within the plane of collimator focus. If it is nearer or beyond the plane of focus it is seen less clearly. If the region in question is too far a distance from the focal plane it may be missed. The greater the diameter of the crystal and collimator, the thinner is the plane of focus and the greater is the tomographic effect. Thus a rectilinear scanner with a 5-inch (diameter) crystal exhibits a shallower focal plane than does one with a three-inch crystal.

The standard gamma camera does not exhibit a tomographic effect. Maximum resolution is obtained at the surface of the collimator. Resolution decreases slightly as the plane of interest is moved away from the collimator surface. A decrease in resolution of approximately 15 percent may be expected at a distance 4 inches beyond the collimator. One firm makes an attachment for its gamma camera that permits tomography to be done.[5] The attachment consists of a special rotating collimator with slanting parallel holes and a table whose top moves in a circular motion on a 4.5-cm radius. The collimator moves at the rate of 2 revolutions per minute (rpm) while the table is simultaneously passing through two cycles of motion. Since the table's motion is limited and it moves slowly the patient experiences no discomfort. Four tomographic scans are generated simultaneously with this device, each showing a different plane of focus. A modification of the image oscilloscope allows all four views to be recorded at once. The tomographic equipment may be easily positioned or removed to allow either standard or tomographic operation. The theory of this device is rather involved and beyond the scope of this book.[6]

The image obtained from a rectilinear scanner is a full size image that has a 1:1 ratio with the anatomy of the patient. One maker of rectilinear scanners provides a means for recording the image in full or reduced size. The reduced image size permits a whole body scan to be recorded on a standard 14 by 17 inch x-ray film. Images from the gamma camera are recorded in much reduced size with an oscilloscope camera that uses film varying from 35 mm to 4 by 5 inches.

An advantage for the rectilinear scanner is the large area that may be included in the scan. A 14- by 17-inch field of observation is more or less standard for rectilinear scanners. Gamma cameras are now limited to a 10- to 12-inch diameter circle. A slightly larger field diameter may be obtained by using a collimator with divergent rather than parallel holes. The divergent collimator produces distortion in the image, however, limiting the value of the increased scanning area.

A single hole collimator may be substituted for the multiple parallel hole collimator to increase or decrease the field covered by a gamma camera. The single hole collimation device consists of a funnel shaped cone placed with the large end against and covering the crystal. Typically the small end of the cone is 20 cm from the crystal and contains a tungsten plug with a hole 0.5 cm in diameter. This single hole collimator operates as a pinhole camera because gamma radiation travels in straight lines. An inverted image of the radiation-emitting area is projected into the crystal (Fig. 8). The projected image may be made to cover larger or smaller areas according to the formula:

$$\frac{\text{Object size (at patient)}}{\text{Image size (at crystal)}} = \frac{\text{object distance (from pinhole)}}{\text{image distance (from pinhole)}}$$

The great speed of the gamma camera is lost with the pinhole collimator, and it becomes comparable to the rectilinear scanner in the time needed to produce an image. Speed may be increased by making the pinhole larger, but then resolution is lost. The pinhole collimator is most often used to obtain a sufficiently large image of the thyroid.

The larger the holes in a collimator, the greater is the sensitivity and the lower is the resolution of the radiation detector system. This rule applies to both rectilinear scanners and gamma cameras. A compromise is usually made between these factors when a collimator is designed. In practice, the operator may have a selection of several collimators that can be interchanged to meet various needs. If the holes are made too large in a parallel hole collimator for a gamma camera, an image of the collimator may appear as an artifact in the scan. Such artifacts can be prevented by setting the collimator into uniform circular motion during the scanning period. In this fashion the

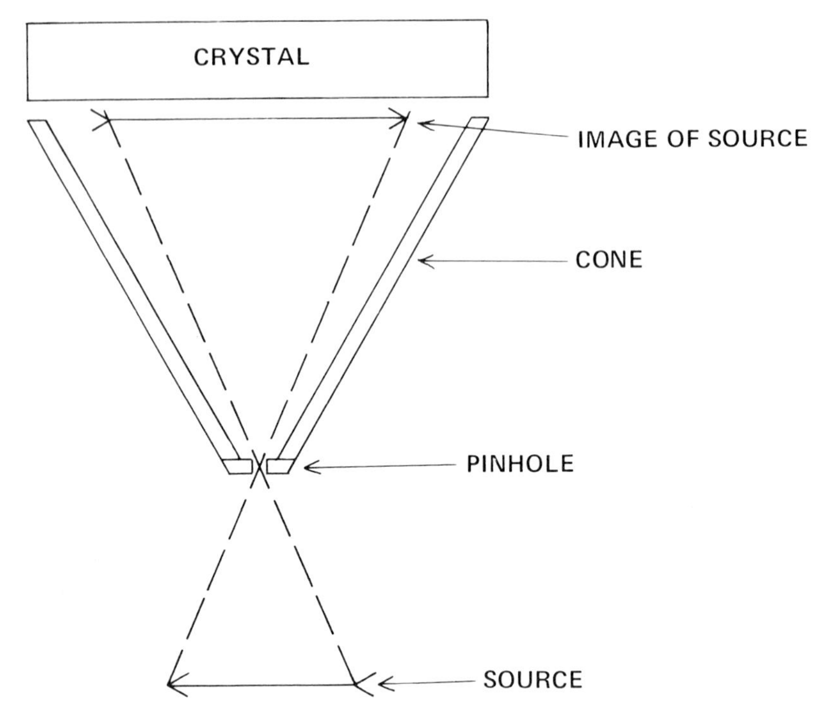

Fig. 8. A pinhole collimator can be used to increase or decrease the area covered by the image formed in a gamma camera.

collimator artifact is blurred out of the scan without deterioration of the image.[7]

The rectilinear scanner has a uniform sensitivity over its entire field of view because it has only one photomultiplier tube. Uniform sensitivity in the gamma camera is attained when all 19 photomultiplier tubes exhibit equal sensitivity. To achieve such equal sensitivity considerable adjustment may be required. Frequent tests must be run to make sure the gamma camera remains properly adjusted.

A rectilinear scanner costs one-half to two-thirds as much as a gamma camera. Many institutions start with a rectilinear scanner because of this. A gamma camera is added when the work load passes beyond the capacity of a single rectilinear scanner. In general the cost of scanning apparatus is comparable to that of x-ray equipment. A gamma camera with a tomographic attachment costs as much as the most sophisticated x-ray machine. Considering the high cost of equipment and the poor quality of scanner images, the question is sometimes raised as to the value of scanning. Compared to a radiograph the scan makes a very poor showing. The resolution of a scan image is about 5 mm compared to a resolution of a fraction of a millimeter for radiographs. However, scans should not be compared to radiographs. Basically the radiograph provides anatomic and the scan physiologic information. Despite the poor image quality, much funtional information is obtained by radionuclide scanning that can not be obtained by other means. Efforts are continuously being made to improve scanner performance. Computer analysis of the images obtained by scanning is one recent approach.[8]

Radiopharmaceuticals for Scanning

In considering scanning, attention must be directed to the radioactive materials administered to the patient. The amount of radioactivity given for organ scanning must be much greater than the amount necessary for functional studies. A thyroid scan requires the administration of about 10 times the minimum activity needed for a thyroid uptake test. The large radiation dosage needed with some radionuclides makes them unsuitable for scanning, while other nuclides are particularly desirable scanning agents. In selecting a nuclide several factors should be considered.

The preferred nuclide should emit only gamma rays. The emission of beta particles contributes heavily to radiation dosage of the patient without participating in image formation. In the case of iodine 131 about 90 percent of the radiation dose received by the patient is due to beta particles. If the iodine 131-containing thyroid of a patient is scanned, the image is formed by gamma radiation only with no contribution from the beta emission. In general, a nuclide that decays by electron capture is preferable because it often emits only gamma radiation.

The effective half-life within the patient should be short. Rapid decay or rapid turnover of the nuclide limits the accumulation of radiation exposure. A short physical half-life is desirable as it automatically limits patient exposure. Disposal of excess activity is also less complicated with short-lived nuclides. In some cases a material of longer physical half-life may be used if it has a short biologic half-life.

The optimum energy range for scanning is 0.10 to 0.20 MeV. The scintillation crystals used in image systems respond with high efficiency in this energy range. Photon energies above 0.20 MeV tend to penetrate the septa of collimators (unless they are extra thick) and degrade the image. High energy photons also have a low efficiency in imparting energy to the detecting crystal. Below 0.10 MeV, radiation absorption within the patient is excessive. Photons emitted deep within the patient do not reach the surface to contribute to the image.

A unique radionuclide with all the desirable properties for scanning is metastable technetium 99 (^{99m}Tc).[9,10] The *metastable state* is one in which an atomic nucleus remains excited for a relatively long period of time before it releases its gamma ray energy. This is best understood by comparing a radioactive decay process that does not involve a metastable state with one that does. In the decay of radioactive cobalt 60 a nuclear neutron changes to a proton by throwing out a beta particle; the result is nickel 60. The new nickel 60 atom promptly emits two gamma rays. Thus the events starting with the cobalt 60 decay are immediately concluded. In the case of radioactive molybdenum 99 a neutron also changes to a proton by throwing out a beta particle, the result being technetium 99. The technetium 99 atoms are in an excited state with excess energy that gives rise to gamma rays. It may be a number of hours before the excess energy is released as gamma radiation from any particular atom. This period of existence in a high energy state is sufficiently long to allow quantities of metastable technetium 99 to be isolated for use as a scanning agent. In summary, technetium 99 may exist in a high energy form (^{99m}Tc) and a low energy form (^{99}Tc). These two forms are said to be isomeric. *Isomerism* exists when there is a relationship between two nuclides such that they have the same mass numbers and atomic numbers but different energy states. Technetium 99m decays to technetium 99 by isomeric transition. In *isomeric transition* the nucleus of an atom changes energy states without any nuclear reorganization. Thus ^{99m}Tc goes to ^{99}Tc with the emission of a 0.14 MeV gamma ray. There is no beta emission. Beta particles come only from the molybdenum from which the technetium 99m was derived. The half-life of metastable technetium 99 is 6 hours.

The Technetium Generator

In the clinical laboratory technetium 99m is usually obtained from a *technetium generator.* This generator consists of an alumina ceramic column with radioactive molybdenum 99 absorbed to its surface as ammonium molybdate. The molybdenum column is enclosed in a glass container with Millipore-filtered entrance ports at the top and bottom. The assembly is autoclaved for sterility and placed in a lead container to provide radiation shielding. Technetium 99m is obtained by pouring an eluting solution through the generator. The process is represented schematically with appropriate decay schemes in Figure 9. The eluting solution washes out the ^{99m}Tc, leaving the ^{99}Mo behind on the ceramic column. The solution coming from the generator contains ^{99m}Tc as sodium pertechnetate (Na $^{99m}TcO_4$). The TcO_4^- ion is very similar to the iodide ion (I^-) and is capable of forming several useful scanning agents. Sodium pertechnetate is used for scanning the brain and thyroid; technetium sulfur colloid for the liver, bone marrow, and spleen; and technetium-labeled microspheres for the lung.

When a technetium generator is used there must be available suitable

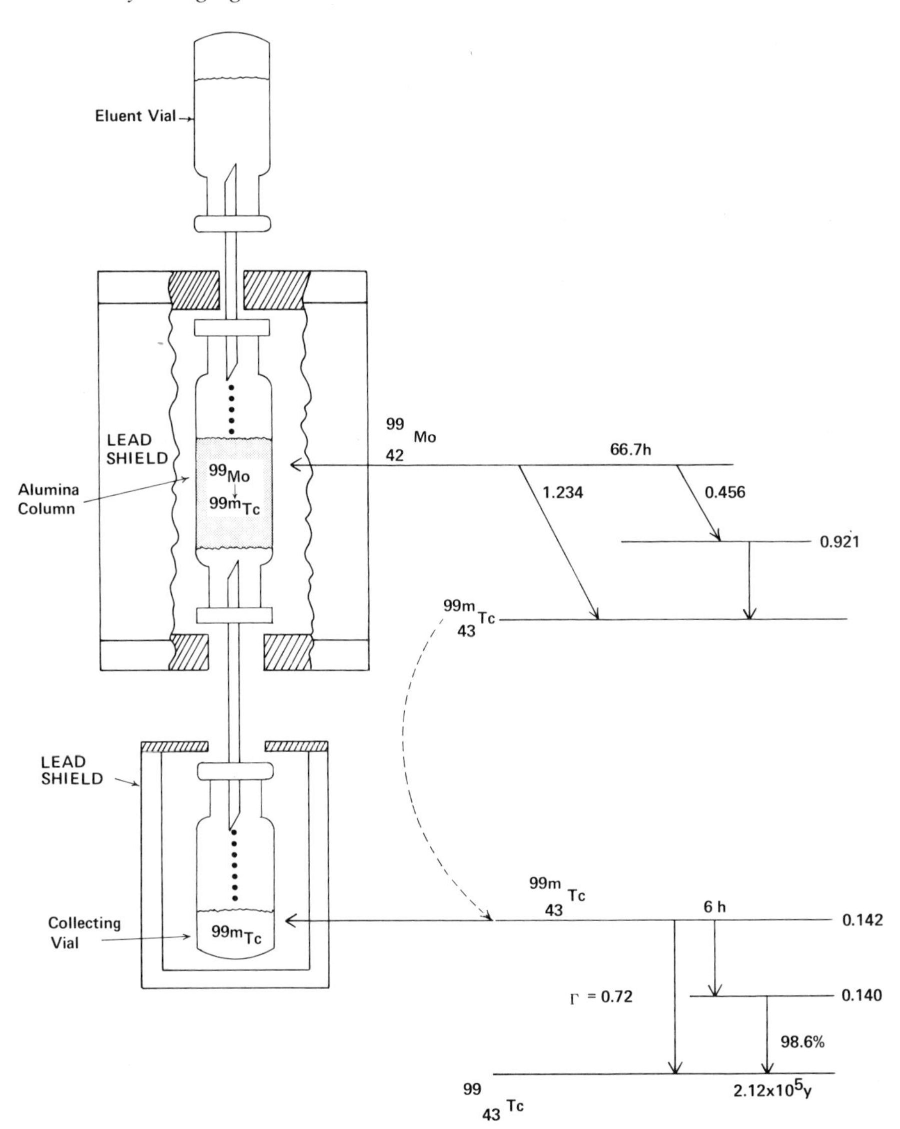

Fig. 9. Liquid flow through a technetium 99m generator during elution. Corresponding decay schemes for the generator and the eluate are given at the right.

equipment, preferably an ionization chamber, to conduct radioassays on the eluate from the generator. The activity of the eluate in terms of millicuries per milliliter must be known to prepare patient dosages. A test for molybdenum 99 must also be made. Molybdenum 99 may break away from a faulty column or be washed out with eluting fluid of improper pH. Eluted technetium 99 must contain no more than 1 μCi ^{99}Mo per millicurie of ^{99m}Tc or 5 μCi ^{99m}Mo per dose of ^{99m}Tc administered.[9,10]

The question arises as to how often a technetium generator can be eluted. The answer to this question resides in the behavior of the parent-daughter relationship between molybdenum 99 and technetium 99m. The term parent-daughter is often used in discussing radionuclides where the parent is a radioactive nuclide (^{99}Mo) and the daughter (^{99m}Tc) is the product of the radioactive decay of the parent. In this case the half-life of the parent is 66.7 hours and that of the daughter 6 hours. The relationship between the decay rates is such that if molybdenum 99 is placed in a generator and left untouched technetium 99m builds up in a few hours to a level that depends upon the amount of molybdenum present. The amounts of molybdenum and technetium then both decrease at the rate of molybdenum decay (Fig. 10). The maximum amount of technetium is present 23 hours from the starting time. If the generator is eluted every 24 hours, the pattern of technetium buildup in the generator is as shown in Figure 11. The maximum amount of ^{99}Tc

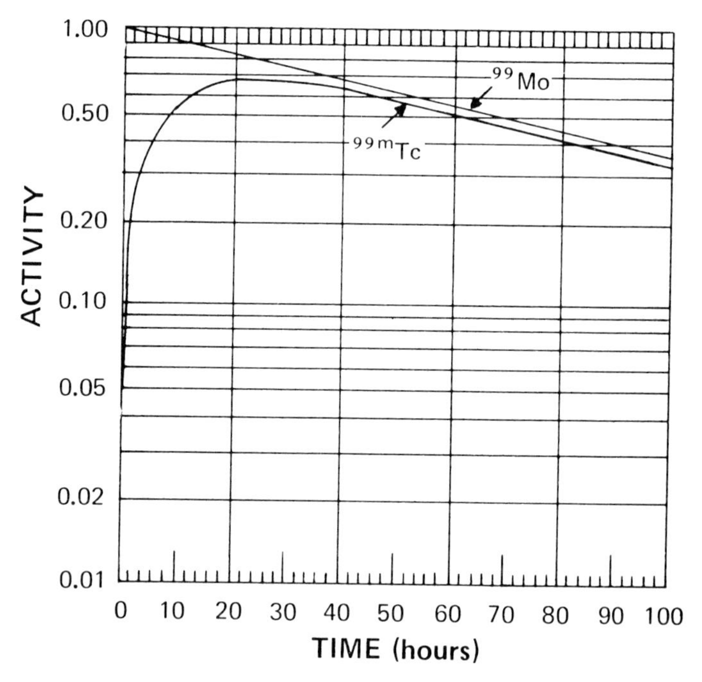

Fig. 10. Growth and decay of ^{99m}Tc in a molybdenum 99 generator. Once the activity of the technetium reaches its maximum the technetium and molybdenum decrease in activity at the same rate.

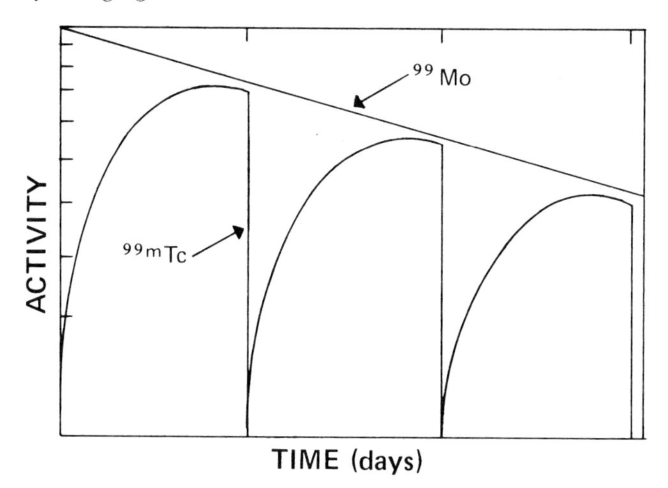

Fig. 11. Decay of molybdenum and growth pattern of technetium associated with daily elution of the generator.

obtainable is obtained every day and is independent of past elutions of the generator, provided the generator was not eluted within the past 23 hours. If the time since the last previous elution is less than 23 hours, the amount obtained is reduced accordingly. As the molybdenum decays the maximum daily amounts become less and less. After 2.78 days (the half-life of molybdenum 99) the amount of technetium 99m obtainable is only half the amount obtainable at the start. A molybdenum generator supplies technetium daily, or more often, for about a week. After a week the specific activity of the eluate is usually below a useful minimum. Users of technetium generators often obtain generators weekly from suppliers on a long-term contract basis.

Other generating systems are possible. One generator has been tried in which radioactive tin 113 decays to metastable indium 113.[11] The parent tin has a half-life of 118 days, and the metastable indium 113 produced by the tin decay has a half-life of 1.73 hours. Indium 113 has been used for liver, lung, and brain scanning. The short half-life of 1.73 hours makes frequent elution of the generator necessary and also limits the time for preparation of the scanning radiopharmaceutical. The whole process of elution to completed scan must be done rapidly. The long life of the parent tin makes generator replacement necessary only after several months. So far the technetium 99m generator remains the only one in wide use.

In the jargon of the laboratory, a radionuclide generator is often called a *cow*, and the process of eluting the desired nuclide, *milking*.

REFERENCES

1. Kenny PJ: Collimation for rectilinear scanners and camera imaging equipment. Semin Nucl Med 3:259, 1973
2. Anger HO: Radioisotope cameras. In Hine GJ (ed): Instrumentation in Nuclear Medicine. New York, Academic Press, 1967, Vol 1, pp 485-552
3. Bender MA, Blau M: Autofluoroscopy: the use of a non-scanning device for localization with radioisotopes. J Nucl Med 1:105, 1960
4. Ter-Pogoaaian NM, Niklas WF, et al: An image tube scintillation camera for use with radioactive isotopes emitting low-energy photons. Radiology 86:466, 1966
5. Muehllehner G: A tomographic scintillation camera. Phys Med Biol 16:87, 1971
6. Freedman GS (ed): Tomographic Imaging in Nuclear Medicine. New York, Society of Nuclear Medicine, 1973
7. Bramlet RC: Ghost images as artifacts in gamma camera scans. Am J Roentgenol Radium Ther Nucl Med 109:676, 1970
8. Brown DW: Quantification of image studies. Semin Nucl Med 3:324, 1973
9. Smith EM: Properties, uses, radiochemical purity and calibration of Tc^{99m}. J Nucl Med 5:871, 1964
10. Smith EM: Internal dose calculations for ^{99m}Tc. J Nucl Med 6:231,1965
11. Penkoske P, Potchen EJ, Welch MJ, Welch TJ: Clinical chemistry of the tin-indium generator. J Nucl Med 10:646, 1969

chapter 11

HANDLING RADIOACTIVE MATERIALS

Since the dawn of time man has been continuously exposed to low levels of ionizing radiation. Internal irradiation arises from naturally occurring radioative elements, such as potassium 40 and carbon 14, which are present in minute amounts within the human body. External irradiation results from cosmic rays and naturally occurring radioactive elements within the earth's crust. Continuous exposure to this internal and external background irradiation seems to have caused no physical damage to man. With the discovery of x-rays and the use of radium certain individuals were subjected to radiation levels many thousands of times greater than background. Among these exposed individuals, local tissue destruction and long-term changes leading to cancer were apparent.[1] Subsequent animal experimentation indicated that ionizing radiations may produce genetic changes which may be carried from generation to generation.[2]

Over the years an extensive study had been made of the biologic effects of radiation. From this accumulated knowledge criteria for the safe use of radiation have been established. Two sets of guidelines have been established: One set pertains to radiation workers and is based upon possible physical damage to the individual. The second set, based on possible genetic damage, pertains to the general public. Patients are excluded from both catagories; radiation exposure of the patient is left to the discretion of the physician, it being assumed that radiation procedures are prescribed only in the best interest of the patient. It is the duty of the physician to see that no patient is exposed to more radiation than is necessary for diagnostic study or treatment.

RADIATION UNITS

Interpretation of the radiation protection guides requires an understanding of the quantitative units association with radiation dosimetry. There are three such units commonly applied to the ionizing radiations.[3] The *roentgen* (R) is a unit for expressing exposure to x or gamma radiation in terms of ionization produced in air. The roentgen specifies the ionization density produced in an irradiated volume of air. It is that amount of ionization in air which is sufficient to produce 1 esu of electrical charge per 0.001293 g of air or 0.000258 coulomb per kilogram of air.

The *roentgen absorbed dose* (rad) is a unit for expressing the radiation dose absorbed in any medium of any kind of radiation. Note that the rad applies to any kind of ionizing radiation, while the roentgen is limited to x or gamma radiation. The roentgen is defined in terms of ion pairs produced, while the rad is defined in terms of absorbed energy. One rad equals 100 ergs of absorbed energy per gram in any medium from any kind of ionizing radiation.

The *roentgen equivalent man* (rem) expresses the estimated equivalent of any type of radiation that would produce the same biologic effect in man as 1 rad delivered by x or gamma radiation. The rem takes into account the fact that living systems respond in different ways to equal amounts of energy delivered by different types of radiation.

The radionuclides used in medicine emit predominantly gamma photons and beta particles of energies such that the roentgen, the rad, and the rem exhibit little difference when used as units of measurement. For purposes of approximate dosage, such as that used in medical radionuclide protection calculations, the roentgen, rad, and rem may be used interchangeably. Commonly used subunits are the milliroentgen, millirad, and millirem. Each subunit is 0.001 of the parent unit. Some feeling may be developed for these units by looking at some known values. Each person accumulates about 20 millirads each year due to radioactive potassium within the body. Total background radiation dose due to internal and external radiation exposure is around 150 to 400 millirads per year, depending upon altitude and nature of surrounding rock formations. A dose to the whole body of 450 rads (450,000 millirads) in one short exposure would be expected to bring about fatality in half of the individuals so exposed. Radiation exposure to the skin from radiographic procedures may vary from less than 20 milliroentgens (mR) for a posterior-anterior chest film to more than 2500 mR for a lateral lumbar spine film.[4] Radiation exposure from diagnostic radioisotope studies is comparable to that for radiographic studies. In certain cases radioisotope studies result in less patient exposure than comparable radiographic procedures.[5] In radiation therapy local areas of the body are often treated with dosages of 5,000 to 7,000 rads. Therapeutic dosages are usually delivered by external beam radiation sources over extended periods of time at the rate of 100 to 200 rads/day.

PERMISSIBLE RADIATION EXPOSURE

The maximum permissible radiation dose to workers in the field is set at a level such that continuous exposure of an individual to that amount of radiation over a long period of time is not expected to cause detectable bodily injury to the person during his lifetime. The maximum permissible radiation exposure to the whole body—including the gonads, blood-forming organs, and lens of the eye—for radiation workers is set at 0.1 rem (100 millirems)/week or 5 rems/year.[6] Radiation workers may receive a maximum of 1.5 rems/week or 75 rems/year to the hands and feet. The maximum accumulated whole body dose for radiation workers, where N is the age of the individual, may be calculated by the formula:

$$\text{Accumulated dose} = 5\,(N - 18)$$

Eighteen is subtracted from the age because persons less than 18 years of age are not employed as radiation workers.

The maximum permissible radiation dose to nonradiation workers (the general public) is set at a level such that continuous exposure of the majority of individuals in the population over a long period of time would not be expected to produce a detectable genetic change in subsequent generations. The maximum weekly exposure to nonradiation workers is 0.01 rems (10 millirems)/week or 0.5 rem/year to the whole body. In practice, a dosage well below the maximum permissible levels can and should be maintained for both radiation workers and the public. Dosage is kept at a low level by minimizing radiation exposure from external sources, by preventing contamination of the body surface with radioactive material, and by preventing internal ingestion of radioactive material.

To protect the public and nonradiation workers radioactive working areas are designated controlled areas and only radiation workers and patients are admitted. Occupied areas adjacent to controlled areas are protected from external irradiation by distance from radioactive sources or by shielding the sources. Protection of the nonradiation working public also places strict limitations upon the release of radioactive material into the environment. Radioactive waste, beyond small specified amounts, may not be released into the sewer, placed in the garbage, or incinerated. Small amounts of short-lived radioactive material may be held in storage until the activity falls to an insignificant level, after which the material may be disposed of through the sanitary sewage system. The quantity of radioactivity that may be disposed of through a sanitary sewer is under the regulation of the state or federal agency which grants licenses for the use of radioactive material. The user may determine from the licensing agency under which he operates the maximum permissible amounts of radioactivity that may be dispersed into a given sanitary sewer. All material so released must be dispersible in water. Activity in larger amounts or of longer life is best disposed of through the service of commercial firms specializing in the disposal of radioactive materials.

RADIATION PROTECTION MEASURES

The radiation worker is largely dependent for protection upon his knowledge of and willingness to use certain techniques. All persons working in controlled areas must wear film badges so that radiation exposure may be continuously monitored. Records of individual cumulative dosage must be maintained. Any indication of excessive dosage should be investigated and the causes eliminated. Limitation of radiation dose from external exposure to gamma-emitting sources involves three factors. The first factor is time. The radiation dose that one receives in the vicinity of a given source is directly proportional to the time spent in the vicinity of the source. Procedures involving radioactive materials should be planned for minimum working time with a minimum number of persons involved. New procedures should be executed with nonradioactive materials until sufficient experience has been gained to make them run smoothly and without delay. Equipment should be kept in good order and materials should be immediately at hand so that delays do not develop during the handling of radioactive substances. On the other hand, work should not be rushed. An attempt to work rapidly often results in spills or other accidents that increase exposure time and impose the risk of contamination. Needless to say, exposure time must not be increased by loitering in the vicinity of radioactive materials. The second factor is distance. Most radiation sources are bottles, vials, or test tubes of small dimensions. The radiation dose rate in the vicinity of such a source varies inversely as the square of the distance from the source.[7] Suppose a vial contains a gamma-emitting material of a quantity such that the dose rate at 10 cm from the vial (distance 1) is equal to 100 mR/hour (dose rate 1). Then the dose rate (dose rate 2) at 20 cm distance (distance 2) is equal to:

$$\text{Dose rate 2} = \text{Dose rate 1} \times \frac{(\text{distance 1})^2}{(\text{distance 2})^2} = 100\ \text{mR/hour} \times \frac{(10\ \text{cm})^2}{(20\ \text{cm})^2}$$

$$= 100 \times \frac{100}{400} = 25\ \text{mR/hour}$$

If the distance is increased from 10 to 100 cm the dose rate falls from 100 mR/hour to 1 mR/hour:

$$\text{Dose rate 2} = 100\ \text{mR/hour} \times \frac{(10\ \text{cm})^2}{(100\ \text{cm})^2} = 100 \times \frac{100}{10{,}000} = 1\ \text{mR/hour}$$

Tables of gamma ray dose rates per hour for 1 mCi of various isotopes at a distance of 1 cm have been made up to assist in calculating dose rates at various distances. The dose rate in roentgens per millicurie per hour at a 1 cm distance is designated the *gamma ray dose rate constant* and is indicated by Γ. For radioiodine 131 this figure is 2.2 R/mCi/hour/cm. Consider a 100/μCi capsule of ^{131}I. What is the radiation dose rate at 10 cm from the capsule? Remember that 1 mCi equals 1000 μCi. Hence:

$$100\ \mu\text{Ci} = \frac{100\ \mu\text{Ci}}{1000\ \mu\text{Ci/mCi}} = 0.1\ \text{mCi}$$

The dose rate at 1 cm from the capsule is:

$$2.2\ \text{R/mCi/hour} \times 0.1\ \text{mCi} = 0.22\ \text{R/hour}$$

At 10 cm the dose rate is:

$$0.22\ \text{R} \times \frac{(1\ \text{cm})^2}{(10\ \text{cm})^2} = 0.22 \times \frac{1}{100} = 0.0022\ \text{R/hour}$$

$$0.0022 \times 1000\ \text{mR/hour} = 2.2\ \text{mR/hour}$$

This type of calculation is valid whenever the distance from the source to the point in question is at least five times the longest dimension of the source.

In practice a safe distance from sources may be achieved by a number of methods: Sources are handled with long forceps or tongs. Shipping containers are of large dimensions with the source supported at the center of the container. Work benches can be organized so that source containers are kept at a maximum distance from the worker. Storage areas are well away from personnel, radiation-counting equipment, and sensitive materials such as film. It is advisable to have the rooms containing radiation detection equipment separate and at some distance from the room where radioactive materials are stored, prepared for dispensation to patients, or prepared for counting.

The third factor used in limiting exposure from external sources is shielding. Here a radiation-absorbing barrier is placed between the radiation source and the worker or protected area. The most commonly used absorbing material is lead. The absorption of gamma radiation in lead, or any other material, follows a mathematical law similar to the law previously described for radioactive decay. For example, let a source of radioiodine 131 be set 10 cm away from a detector with a ratemeter. Assume an activity such that the ratemeter indicates a radiation level of 100 mR/hour. If a sheet of lead 3.3 mm thick is placed between the source and the detector the ratemeter reading falls to one-half the reading without the lead, or 50 mR/hour. If a second 3.3 mm thickness of lead is placed between the source and the detector, half the radiation emerging from the first sheet of lead is removed and the meter reading drops to 25 mR/hour. If a third sheet of lead is added the reading is 12.5 mR/hour. In short, a 3.3 mm thickness of lead reduces the radiation level around any iodine 131 source by 50 percent. This thickness of 3.3 mm is called the *half-value layer* of lead for iodine 131. The half-value layer varies with the material forming the barrier and with the energy of the radiation passing through the barrier. Tables of half-value layers in lead or other material for specific radionuclides may be used to calculate necessary thickness for radiation barriers. When calculating, it helps to remember that seven half-value layers between a radiation source and a point reduce the radiation level at the point to 1 percent of that without the barrier. Ten half-value layers reduce the radiation level by a factor of 1000. If the radiation level at a given point is 10 roentgens/hour, the introduction of 10 half-value layers of absorber between the source and the point reduces the level to 10 milliroentgens/hour.

Shielding with lead is often used when shipping and storing radioactive material. Storage areas may be provided with shielding in the form of lead bricks, and lead shields may be used around syringes.

Despite the common use of lead for shielding, the lead apron and lead gloves of the radiologist should not be used in the radionuclide laboratory. These articles contain about 0.25 mm thickness of lead, which is equivalent to four or more half-value layers for diagnostic x-rays. For most radioactive materials 0.25 mm of lead is much less than one half-value layer and the protection provided is insignificant. A lead apron thus provides a false sense of security and only impedes movement.

Internal radiation exposure follows the introduction of radioactive materials into the body. The routes of accidental bodily entry are by swallowing, inhalation, and absorption through the skin. Once a radioactive material enters the bloodstream, control over it is lost. Efforts to speed elimination produce limited results. Most radioactive materials remain in the body to be gradually eliminated by the combined processes of physical decay and biologic turnover. Techniques to prevent accidental introduction of radioactivity into the body are mandatory for the radiation worker. The following work rules greatly reduce the possibility of incidents leading to internal radioactive contamination.

1. Rubber or plastic disposable gloves should be worn when working with unsealed radioactive substances. The gloves protect against skin contamination. No protection is provided against radiation from sources external to the gloves.
2. Needless contamination by handling objects with contaminated gloves should be avoided. Gloves should be removed before handling light switches, water faucets, doorknobs, etc.
3. Contaminated gloves should be washed before removal.
4. No liquid should be pipetted by mouth in a radioisotope laboratory.
5. No food or beverages should be allowed in the working areas.
6. Smoking and the use of cosmetics should be prohibited.
7. Only self-adhesive labels should be used. Avoid labels that must be wetted before use.
8. All operations that might produce radioactive fumes must be conducted in a force ventilated fume hood. A filtered exhaust suitable for trapping radioactivity may be necessary.
9. Gases such as xenon 133 must be handled in a room with suitable forced ventilation. Use should be limited to enclosed systems with suitable gas traps.

The possibility of a radioactive spill is always present in a laboratory handling radioactive material. Spillage always increases external radiation exposure and enhances the danger of internal contamination.[8] Spills may be minimized by carrying radioactive materials in double containers. Glass bottles are often enclosed in lead pots or metal cans. Transfer operations are best carried out with the dispensing and receiving containers in a shallow metal pan. The bottom of the pan should be covered with plastic-backed absorbent paper to prevent the spread of any spilled material.

Every radioisotope laboratory should have a person trained in radiation protection, monitoring, and decontamination to act as radiation safety officer. Any major spillage of radioactivity should be handled by this officer. A major spill consists of an amount of radioactivity equivalent to or greater than a therapeutic dose. An amount of radioactivity equal to or less than a diagnostic dose may be classified as a minor spill. If a major or minor spill occurs immediate action must be taken to confine the area of contamination. In the case of liquids, dry absorbent material such as paper toweling should be dropped into the liquid to prevent spreading. In the case of solids a moist absorbent material such as dampened paper toweling should be dropped over the spill to keep dust from flying.

Minor spills may be cleaned up by collecting as much of the spilled activity as possible on absorbent material. The area may then be scrubbed with soap or detergent and water. Commercial decontaminants that contain complexing agents to hold radioactivity in solution may be helpful. In removing activity from a contaminated area, work from the sides of the spill toward the center to avoid spreading contamination. Brushes should not be used as the bristles splatter the material, thus spreading contamination. All toweling and other material used in decontamination should be collected in a plastic bag and held for radioactive decay or disposal. Contamination is best removed from the skin by repeated washing with mild soap and water; care should be used not to break the skin. Contaminated clothing should be removed and placed in a plastic bag before leaving the area. Clothing may be sent to the laundry after it has been stored for sufficient time for the radioactivity to decay. The progress of decontamination should be monitored with a portable radiation survey meter. Efforts should be continued until a background reading is obtained or until repeated scrubbing yields no further reduction in radiation level.

In case of a major spill absorbent material should be dropped to confine the area of contamination. All personnel should leave the area. If there is danger of fumes or flying dust, fans, ventilators, and air-conditioners should be turned off. Windows and doors should be closed. The room should be locked or the area roped off and marked with radiation warning signs until the arrival of the radiation safety officer. Any individual whose body or clothing may be contaminated should remain in the vicinity until he has been surveyed by the safety officer. The room should remain unoccupied, except for those persons involved in decontamination, until declared safe by the radiation safety officer. The officer should always be notified of major spills, and the decontamination process should be supervised by him.

One final and important point should be stressed concerning contamination. Great care must be used to prevent contamination of counting equipment. A single drop of radioactivity on the external surface of a test tube can contaminate a well counter to the point where background becomes intolerable. A test tube broken in a well can incapacitate the well for weeks. Spillage of radioactivity into the collimator of an imaging system may disrupt the system for a long time. A portable survey meter is apt to become useless the moment it touches a heavily contaminated object.

The administration of radioactive drugs to patients for diagnosis or treatment requires a balancing of benefit against risk. The benefit is usually obvious, the risk elusive. Ionizing radiation is known to produce both genetic and somatic effects.[9] The mutation rate in a population increases in a manner proportional to radiation exposure. Somatic effects may be immediate and acute or delayed and chronic. The therapeutic use of radioactivity usually produces acute effects within a few days to a few weeks. The diagnostic use of radioactive materials is associated with radiation levels below those necessary for acute reactions. Both therapeutic and diagnostic uses of radioactivity may contribute to the induction of cancer or leukemia—often after a latent period of many years. There is evidence that radiation exposure contributes to a shortened life span by accelerating the aging process.[10]

The rapidly growing tissues in children are more sensitive to radiation than adult structures. The ability of radiation to induce teratogenesis is of particular importance.[11] Malformation has been reportedly induced in mammals with as little as 2 rads applied to the embryo in utero.

The importance of risk in patients may be largely scaled according to age. The patient beyond reproductive age exhibits no genetic liability and presents minimum probability for the manifestation of delayed somatic effects. Diagnostic radionuclides may be used in patients over 50 years of age without great concern. Between the age of 25 and 50 years the use of radionuclides should be reserved for specific needs. These patients are a genetic liability and have an expected life span long enough to exhibit delayed effects. In those under 25 years of age the use of radioactivity is best reserved for serious medical indications. As a group these patients exhibit the greatest genetic liability; they have an expected life span of many years plus increased radiosensitivity in their still-growing tissues.

The use of radioactivity in a pregnant female should be avoided if possible. On the basis of data transposed from rodents and primates, the human fetus is most vulnerable to radiation at 32 to 37 days after conception.[12] The fetus should be protected from radiation, especially during the first 6 weeks of gestation, in order to preclude radiation-induced malformation. The National Committee on Radiation Protection advises that radiation exposure during the entire period of pregnancy be limited to 0.5 R or less for a radiation worker.[13] In Denmark it has been proposed that therapeutic abortion be carried out if a fetus is exposed to 10 R or more during the first 4 months.[14] Fetal doses between 1 and 10 during the first 4 months are considered cause for abortion if there are additional indications. Fetal dosages below 1 R are not considered of sufficient danger to make abortion a consideration.[15] No specific rules concerning the radiation of fetal life, except for the 0.5 R rule for radiation workers, have as yet been endorsed by any of the regulatory agencies in the United States.

REFERENCES

1. Brecher R, Brecher E: The Rays; A History of Radiology in the United States and Canada. Baltimore, Williams & Wilkins, 1959, pp 161-174
2. Gaulden ME: Genetic effects of radiation. In Dalrymple GV, Vogel HH, Gaulden ME, Kollmorgen GM (eds): Medical Radiation Biology. Philadelphia, Saunders, 1973, pp 52-83
3. Leucutia L: More on the r and c units and some other related units. Amer J Roentgenol Radium Ther Nucl Med 117:712, 1973
4. Antoku S, Russell WJ: Dose to active bone marrow, gonads and skin from roentgenography and fluoroscopy. Radiology 101:669, 1971
5. Witcofski RL: Radiobiologic aspects of nuclear medicine. In Dalrymple GV, Gaulden ME, Kollmorgan GM, Vogel HH (eds): Medical Radiation Biology. Philadelphia, Saunders, 1973, pp 246-249
6. NCRP Report 39: Basic Radiation Protection Criteria. Washington, DC, National Council on Radiation Protection and Measurements, 1971, pp 106-107
7. Quimby EH, Feitelberg S, Gross W: Radioactive Nuclides in Medicine and Biology, 3rd ed. Philadelphia, Lea & Febiger, 1968, pp 162-166
8. NCRP Report 8: Control and Removal of Radioactive Contamination in Laboratories. Washington, DC, National Council on Radiation Protection and Measurements, 1951
9. Bushong SC: Radiation dose-response relationships and human disease. In: Medical Radiation Information for Litigation. Rockville, Md, US Public Health Service Bureau of Radiological Health, 1969, pp 60-81
10. Vogel HH: Radiation life shortening. In Dalrymple GV, Gaulden ME, Kollmorgan GM, Vogel HH (eds): Medical Radiation Biology. Philadelphia, Saunders, 1973, pp 225-229
11. Russell LB, Russell WL: Pathways of radiation effects in the mother and embryo. Cold Spring Harbor Symp Quant Biol 19:50, 1954
12. Roberts R: X-ray-induced teratogenesis in the mouse and its possible significance to man. Radiology 99:433, 1971
13. NCRP Report 39: Basic Radiation Protection Criteria. Washington, DC, National Council on Radiation Protection and Measurements, 1971, pp 92-93
14. Hammer-Jacobsen E: Therapeutic abortion on account of x-ray examination during pregnancy. Dan Med Bull 6:113, 1959
15. Brent RL, Spar IL, Jacobsen IV: Radiation exposure during first trimester: when is abortion indicated? JAMA 224:536, 1973

APPENDIX

Data on Radionuclides Commonly Used in Nuclear Medicine Imaging (Normal Subjects)

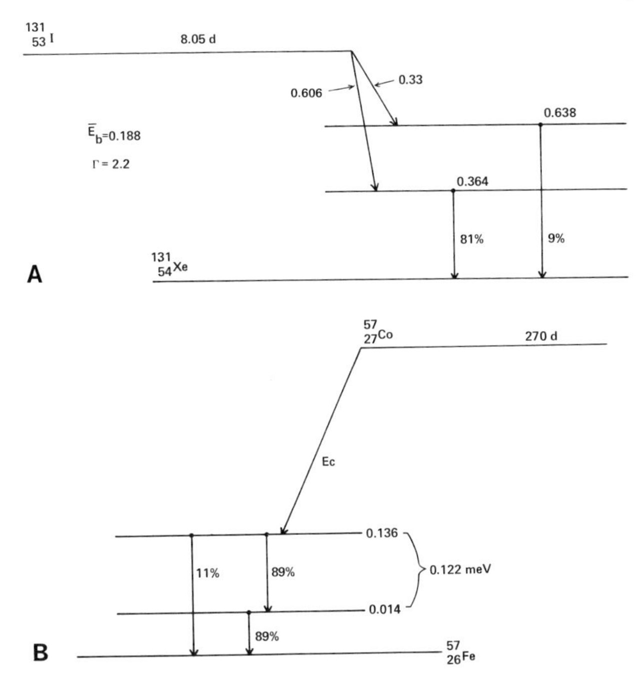

Examples of decay schemes. A. Two major routes for the decay of iodine 131. B. Complete scheme for the single mode of decay of cobalt 57 (see Figure 4, p. 197).

NOTES:

1. Major modes of decay are identified as follows:
   EC = electron capture β^- = decay by beta emission
   IM = isomeric transition β^+ = decay by positron emission
2. Major gamma energies are specified in MeV (millions of electron volts). IC = internal conversion (See item 5).
3. Gamma ray dose rate constant (Γ) is specified in R pr per mCi/hr at 1 cm distance.
4. HVL specifies half value layer in mm of lead.
5. $\bar{E}_\beta$ = energy of average beta emission in MeV.
   (A) Values for $\bar{E}_\beta$ are found for some nonbeta emitting nuclides. These nuclides emit gamma or x-rays of less than 11.3 keV (0.0113 MeV). Such low energy photons travel short distances in tissue, compared to beta particles; therefore, they are included in beta dosage calculations rather than in gamma dosage figures.
   (B) Some low energy gamma photons may undergo IC within an atom so that an orbital electron is ejected from the atom in place of the gamma photon. These electrons contribute to the beta dosage.

APPENDI

DATA ON RADIONUCLIDES COMMONL

(NORMA

Isotope	Major mode of decay	Major Gamma energies	$T_{1/2}$ physical	$\bar{E}_\beta$	Γ	HVL Pb (mm)	Radiopharmaceutical	Use	Usual adult dos
Chromium 51 (^{51}Cr)	EC	0.323 (8%)	27.8 days	0.005	0.15	2.0	^{51}Cr-Sodium chromate	Spleen imaging	300 μCi
Fluorine 18 (^{18}F)	β^+	0.510 (194%)	1.8 hr	0.279	4.4	4.0	^{18}F-Sodium fluoride	Bone imaging	2 mCi
Gallium 67 (^{67}Ga)	EC	0.093 (68%) 0.184 (23%) 0.269 (21%)	3.25 days	0.009	1.0	1.5	^{67}Ga-Gallium citrate	Tumor, abscess imaging	2 mCi
Gold 198 (^{198}Au)	β^-	0.411 (95%)	2.7 days	0.315	2.3	3.0	^{198}Au-Colloidal gold	Liver imaging	150–300 μ
Indium 111 (^{111}In)	EC	0.173 (89%) 0.247 (94%)	2.8 days		2.2	2.0	^{111}In-DTPA	Cisternography	500 μCi
							^{111}In-Indium chloride	Tumor seeking agent	1–2 mC
								Hematopoietic marrow imaging	1–2 mC
Indium 113m (^{113m}In)	IM	0.390 (100%) (35% IC)	1.7 hr	0.110	1.75	3.0	^{113m}In-Indium chloride	Imaging for pericardial effusions	2 mCi
								Placental imaging	1 mCi
Iodine 131 (^{131}I)	β^-	0.364 (80%) 0.637 (9%)	8.04 days	0.180	2.2	3.1	^{131}I-Sodium iodide	Thyroid imaging	25–50 μ
								Imaging of functioning thyroid metastases	2 mCi (after TS
							^{131}I-Macroaggregated serum albumin (^{131}I-MAA)	Perfusion lung imaging	300 μC
							^{131}I-Orthoiodohippurate (^{131}I-Hippuran)	Renal imaging	200–40
							^{131}I-Human serum albumin (^{131}I-HSA)	Cisternography	100 μ
							^{131}I-Rose bengal	Liver imaging	150–30
Mercury 197 (^{197}Hg)	EC	0.077 (99%)	2.7 days	0.079	0.31	0.1	^{197}Hg-Chlormerodrin	Renal imaging	200 μ
Mercury 203 (^{203}Hg)	β^-	0.279 (100%)	47 days	0.100	1.2	2.0	^{203}Hg-Chlormerodrin	Renal imaging	100 μ
Selenium 75 (^{75}Se)	EC	0.138 (24%) 0.269 (71%)	120 days	0.019	1.84	2.0	^{75}Se-Selenomethionine	Pancreatic imaging	250 μ
								Parathyroid adenoma imaging	250 μ
Strontium 85 (^{85}Sr)	EC	0.510 (99%)	64 days	0.014	3.0	4.0	^{85}Sr-Strontium chloride or nitrate	Bone imaging	100 μ

SED IN NUCLEAR MEDICINE IMAGING
JBJECTS)

When dministered fore imaging	Localization	Preferred imaging device (stationary or moving detector)	Critical organ	Radiation dose to critical organ (rads)	Total body radiation dose (rads)
3–24 hr	Tagged heat damaged red cell, sequestered in splenic pulp	Moving	Spleen	10.0	0.12
3 hr	Hydroxyapatite crystal of reactive bone formation	Moving	Bone	0.6	0.1
72 hr	Accumulates where cells are undergoing rapid division	Stationary or moving	Bone marrow	3.0	0.6
1–24 hr	Colloidal particles phagocytized in Kupffer cells	Moving	Liver	6.4–12.8	0.35–0.70
2, 6, 24 hr	Subarachnoid space	Stationary or moving	Brain	0.8	0.04
			Spinal cord	0.2	
			Vertebral bone marrow	0.15	
48–72 hr	Unknown; circulates as ^{111}In-transferrin complex	Stationary or moving	Liver	4.5–9.0	0.5–1.0
24–72 hr	Circulates as ^{111}In-transferrin complex, enters RE cells like Fe-transferrin	Stationary or moving	Bone marrow	2.4–4.8	0.5–1.0
			Liver	4.5–9.0	
15 min	Vascular; circulates as ^{113m}In-transferrin complex	Moving	Blood	0.292	0.034
15 min	Vascular, maternal circulation only: circulates as ^{113m}I-transferrin complex	Moving	Maternal blood	0.146	0.017 (maternal)
			Fetal blood	0.008	
24 hr	Becomes trapped in follicular cells and then organified in thyroid hormone	Stationary or moving	Thyroid	50–100	0.088–0.175
, 48, 72 hr	Becomes trapped in functioning tumor follicular cells	Stationary or moving	Whole body (presume thyroid ablated)	0.26	0.26
?–5 min	Particles of 25–75μ diameter become lodged in pulmonary capillaries and precapillary arterioles	Stationary or moving	Lungs	1.9	0.008
–30 min	Secreted in cells of proximal renal tubule	Stationary or moving	Thyroid	With prior KI po 3.8–7.6	0.032–0.064
				Without KI po 26–52	
			Bladder	1.56–3.12	
			Kidneys	0.014–0.028	
, 6, 24 hr	Subarachnoid space	Stationary or moving	Spinal cord	7.24	0.28
15 min	Hepatic parenchymal cells	Stationary or moving	Liver	0.5–0.9	0.15–0.3
?–24 hr	Renal tubule cells	Stationary or moving	Kidneys	0.874	0.002
–24 hr	Renal tubule cells	Moving	Kidneys	6.5	0.150
5 min	Polypeptide hormone incorporation	Stationary or moving	Kidney	11.0	2.0
			Pancreas	3.0	
5 min	Polypeptide hormone incorporation (must suppress thyroid uptake of ^{75}Se)	Stationary or moving	Kidneys	11.0	2.0
72 hr	Hydroxyapatite crystal of reactive bone formation	Moving	Bone	5.6	0.652

Isotope	Major mode of decay	Major Gamma energies	$T_{1/2}$ physical	$\bar{E}_\beta$	Γ	HVL Pb (mm)	Radiopharmaceutical	Use	Usual adult dos
Technetium 99m (^{99m}Tc)	IM	0.140 (94.4%) 0.142 (1.6%) (9.5% IC)	6.04 hr	0.014	0.70	0.3	^{99m}Tc-Sodium Pertechnetate	Brain imaging	10–15 mC
								Radionuclide angiography or venography	10–15 mC
								Thyroid imaging	1–2 mC
							^{99m}Tc-Sulfur Colloid	Liver and spleen imaging	1–2 mC
								RES marrow imaging	1–3 mC
							^{99m}Tc-DTPA (Sn)	Renal imaging	5–10 m
								Brain imaging	10–15 m
							^{99m}Tc-Iron-ascorbate-DTPA	Renal imaging	5 mCi
							^{99m}Tc-Human serum albumin	Placental imaging	1 mCi
								Radionuclide angiography (esp. for pericardial effusions)	1–4 m
								Cisternography	2 mC
							^{99m}Tc-Human serum albumin microspheres and ^{99m}Tc – MAA	Perfusion lung imaging	3 mC
							^{99m}Tc-Polyphosphates; Diphosphonate; Pyrophosphates	Bone imaging	10 mC
Xenon 133 (^{133}Xe)	β^-	0.081 (99%)	5.3 days	0.12	0.73	0.1	^{133}Xe gas	Pulmonary ventilation imaging	10–20
							^{133}Xe dissolved in saline	Pulmonary perfusion imaging	10 m
								Pulmonary ventilation imaging	10 m
Ytterbium 169 (^{169}Yb)	EC	0.177 (21%) 0.198 (33%)	32 days	0.01	1.95	1.0	^{169}Yb-DTPA	Cisternography	1 mC

When dministered fore imaging	Localization	Preferred imaging device (stationary or moving detector)	Critical organ	Radiation dose to critical organ (rads)	Total body radiation dose (rads)
At time of imaging: A-V nalformations	Vascular	Stationary	Colon	1.0–1.4	0.12–0.18
One hour: nost tumors, arcts, subdural hematomas	*Meningiomas:* probably vascular spaces, extracellular *Tumors of brain parenchyma and those of nerve sheath origin:* probably in both tumor cells and extracellular space *Nonneoplastic space occupying lesions (eg, abscess):* probably vascular *Vascular-type lesions (infarcts, subdurals, etc.):* related to the resulting structural alteration	Stationary or moving			
At time of imaging	Vascular	Stationary			
min to 1 hr	Trapped in follicular cells like iodide, not organified	Stationary or moving	Thyroid (no prior $KC10_4$)	0.27–0.54	0.013–0.026
5–30 min	Colloidal particles phagocytized: liver–Kupffer cells; spleen–macrophages	Stationary or moving	Liver	0.32–0.64	0.016–0.032
0–30 min	Bone marrow: RE cells	Stationary	Bone marrow	0.026–0.078	0.016–0.048
			Liver	0.32–0.96	
5–15 min	Renal cortex (excreted exclusively by glomerular filtration)	Stationary	Bladder	2.5–5.0	0.1–0.2
1 hr	Does not enter thyroid, salivary glands, gastric mucosa	Stationary	Bladder	5.0–7.5	0.2–0.3
min to 2 hr	Protein binding of some ^{99m}Tc in renal tubules	Stationary or moving	Bladder	2.8	0.08
5 min	Vascular	Stationary or moving	Maternal: blood	0.050	0.012
			Fetal: blood	0.014	$<$0.010
t time of imaging	Vascular	Stationary	Blood	0.05–0.20	0.012–0.048
, 6, 24 hr	Subarachnoid space	Stationary	Spinal cord	5.4	0.08
5 min	Particles of 10–30μ become lodged in pulmonary capillaries and precapillary arterioles	Stationary	Lungs	1.2	0.0016
2 hr	Hydroxyapatite crystal of reactive bone formation	Stationary or moving	Skeleton	0.45	0.11
ring first min after halation	Respiratory tree	Stationary	Lung	0.050 (10 mCi, single breath and washout)	0.002
			Lung	0.300 (20 mCi, rebreathing 4 min and washout)	0.008
30 sec	Perfused alveolar surfaces (supplied by pulmonary capillaries)	Stationary	Lung	0.04	0.003
athing after pulmonary on imaging	Respiratory tree	Stationary	Lung	0.24 (4 min)	0.012 (4 min)
6, 24 hr	Subarachnoid space	Stationary or moving	Spinal cord	12–14	0.02–0.07
			Brain	1.1	

INDEX

Names of elements are listed alphabetically by standard abbreviation. Pages on which illustrations appear are shown in italics.